Bibliografische Information der Deutschen Nationalbibliothek

Die Deutsche Nationalbibliothek verzeichnet diese Publikation in der Deutschen Nationalbibliografie; detaillierte bibliografische Daten sind im Internet über http://dnb.d-nb.de abrufbar.

ISBN 978-3-8325-4165-1

Logos Verlag Berlin GmbH
Comeniushof, Gubener Str. 47,
10243 Berlin
Tel.: +49 (0)30 42 85 10 90
Fax: +49 (0)30 42 85 10 92
INTERNET: http://www.logos-verlag.de

High Performance Propagation of Large Object Populations in Earth Orbits

Von der Fakultät für Maschinenbau
der Technischen Universität Carolo-Wilhelmina zu Braunschweig

zur Erlangung der Würde

eines Doktor-Ingenieurs (Dr.-Ing.)

genehmigte Dissertation

von: Marek Möckel
aus: Oldenburg (Oldb.)

eingereicht am: 15.09.2015
mündliche Prüfung am: 29.10.2015

Gutachter: Prof. Dr.-Ing. Enrico Stoll
Prof. Kefei Zhang, PhD

2015

I love humans. Always seeing patterns in things that aren't there.

The Doctor

Contents

List of Abbreviations xiii

List of Symbols xvii

Abstract xix

Zusammenfassung [German Abstract] xxi

1. Introduction 1
- 1.1. Space Debris . . . 2
- 1.2. Scope of Work . . . 5
  - 1.2.1. Computational Models for Orbital Propagation . . . 5
  - 1.2.2. Use Cases . . . 5
- 1.3. Outline . . . 8

2. State of the Art 9
- 2.1. Orbital Physics and Propagation . . . 9
  - 2.1.1. Perturbation Forces . . . 10
- 2.2. GPU Computing . . . 18
  - 2.2.1. A Short History of Graphics Processors . . . 18
  - 2.2.2. General Purpose GPU Computing . . . 24
  - 2.2.3. CUDA . . . 25
  - 2.2.4. Alternative Parallel Programming APIs . . . 32
  - 2.2.5. GPU Computing in Space Research . . . 32
- 2.3. Software Architecture and Development . . . 33
  - 2.3.1. Object-Oriented Programming Techniques . . . 33
  - 2.3.2. Design Patterns . . . 35

3. A Software Framework for Orbital Propagators 37
- 3.1. Properties of Orbital Propagators . . . 37
  - 3.1.1. Complexity . . . 37
  - 3.1.2. Modularity . . . 37
  - 3.1.3. Eligibility for Parallelization . . . 37
- 3.2. Orbital Propagation Interface . . . 38
  - 3.2.1. Overview . . . 38
  - 3.2.2. Concept . . . 39
  - 3.2.3. Data Types . . . 43
  - 3.2.4. Host Interface . . . 45
  - 3.2.5. Plugin Interface . . . 46
  - 3.2.6. PropagatorProperties . . . 50
  - 3.2.7. CUDA Support . . . 53
  - 3.2.8. Multi-Language Support . . . 54

3.2.9. Collision Detection 54
3.3. Propagator Implementation Guidelines 55

4. High-Performance Analytical Propagation **61**
4.1. FLORA 61
4.1.1. Overview 61
4.1.2. Atmospheric Model 62
4.1.3. Third Body Perturbations 62
4.1.4. Solar Radiation Pressure 63
4.1.5. Zonal Harmonics 63
4.1.6. FLORA as an OPI Plugin 63
4.2. Ikebana - A Parallel CUDA Propagator 64
4.2.1. Overview 64
4.2.2. Ikebana::Ikebana 65
4.2.3. Ikebana::PMMeanMotion 67
4.2.4. Ikebana::PMZonalHarmonics 69
4.2.5. Ikebana::PMLuniSolar 71
4.2.6. Ikebana::PMSolarRadiation 72
4.2.7. Ikebana::AtmosphericData 74
4.2.8. Ikebana::PMAtmosphere 75

5. Performance Analysis **77**
5.1. Reference Population 77
5.2. Accuracy 78
5.2.1. Floating Point Considerations 78
5.2.2. Accuracy Dertermination 80
5.2.3. Individual Model Accuracy 81
5.2.4. Total Accuracy Results 96
5.2.5. TLE Data Comparison 107
5.3. Speed 109
5.3.1. Runtime Evaluation 109
5.3.2. Benchmarking 110
5.3.3. Performance Evaluation Setup 111
5.3.4. Performance Results 113
5.3.5. CUDA Runtime Analysis 114
5.4. Double Precision Comparison 118
5.5. Summary 126

6. Use Case Study: Space Debris Visualization **127**
6.1. Overview 127
6.2. Classes 128
6.2.1. DOCTOR::DOCTOR 128
6.2.2. DOCTOR::SpaceObject 129
6.2.3. DOCTOR::Debris 130
6.2.4. DOCTOR::TimeMachine 130
6.2.5. DOCTOR::GuiWrapper 130
6.2.6. DOCTOR::ScriptEngine 130
6.2.7. Auxiliary Classes 132
6.3. Propagation 132

6.3.1. GPGPU Approach . . . 132
6.3.2. OPI Approach . . . 133
6.4. Performance . . . 135

7. Conclusions and Further Research **137**
7.1. OPI . . . 137
7.2. Ikebana . . . 138

8. Outlook **141**
8.1. GPU Computing . . . 141
8.2. Numerical Propagation . . . 141

A. CUDA Profiler Report for the Atmospherical Model **143**
A.1. GeForce GTX 860m . . . 143
A.2. GeForce GTX 960 . . . 152
A.3. Tesla K20c . . . 161

B. Individual Error Rate Plots **171**

C. Ikebana Class Headers **177**
C.1. Ikebana::Ikebana . . . 177
C.2. Ikebana::PMMeanMotion . . . 181
C.3. Ikebana::PMZonalHarmonics . . . 182
C.4. Ikebana::PMLuniSolar . . . 183
C.5. Ikebana::PMSolarRadiation . . . 185
C.6. Ikebana::AtmosphericData . . . 187
C.7. Ikebana::PMAtmosphere . . . 190

List of Tables

1.1. Summary of the different propagator use cases and requirements. 8

3.1. Excerpt from [Reglitz, 2012], table 3.18, showing the names of the input parameters that the three orbital propagators FLORA, FOCUS1 and ZUNIEM have in common. 40
3.2. Description of the elements in *OPI::ObjectProperties*. 45
3.3. Constraints for orbital data provided by the host 56
3.4. Constraints for object properties provided by the host 57

4.1. Input values for which NRLMSISE data is provided in FLORA. 62

5.1. Results of the algorithm from listing 5.1. 79
5.2. Classification of the results. 96
5.3. Mean decay rates of FLORA and Ikebana over different propagation times. . . 111
5.4. GPU specifications 112
5.5. CPU specifications 113
5.6. Performance of FLORA on Intel(R) Core(TM) i7-3820 CPU [MP/s] 113
5.7. Performance of FLORA on Intel(R) Core(TM) i7-4710HQ CPU [MP/s] 115
5.8. Performance of Ikebana on NVIDIA Tesla K20c [MP/s] 115
5.9. Performance of Ikebana on NVIDIA GeForce GTX 860M [MP/s] 116
5.10. Performance of Ikebana on NVIDIA GeForce GTX 960 [MP/s] 116
5.11. Performance of Ikebana (Double Precision) on NVIDIA Tesla K20c [MP/s] . . . 119
5.12. Performance of Ikebana (Double Precision) on NVIDIA GeForce GTX 860M [MP/s] 119
5.13. Performance of Ikebana (Double Precision) on NVIDIA GeForce GTX 960 [MP/s] 120

6.1. Speed of DOCTOR with Ikebana, simple CUDA and shader propagation (in frames per second) 136

List of Figures

0.1. Screenshot of DOCTOR showing the wonderful people who made this possible. Thank you so much! xxii

1.1. Illustration of the >1cm object population in Earth orbit as of 2009. 2
1.2. Simulation of a collision between two satellites (a) minutes after the event and (b) six months later, showing the distribution of the generated fragments. . . 4
1.3. A spatial density plot from MASTER 2009 showing various sources of space debris. The region around the 800 kilometre altitude band shows the highest number of objects. 4
1.4. Illustration of the practical difference between analytical and numerical models. 6

2.1. Parameters of an orbital ellipse (based on [Wiedemann, 2014]). 9
2.2. Parameters describing the position of the orbit and the satellite (based on [Wiedemann, 2014]). 11
2.3. Illustration of the eccentric anomaly (based on [Wiedemann, 2014]). 12
2.4. Illustration of Kepler's second law based on [Vallado, 2007]. The grey areas have the same size. 12
2.5. Zonal, sectoral and tesseral harmonics 13
2.6. Visualization of the RAAN change caused by zonal harmonics perturbations. . 14
2.7. Visualization of the apogee decrease caused by atmospheric drag. 15
2.8. Simulation of third body effects causing an inclination buildup in GEO orbits. 16
2.9. Illustration of the simplified shadow model. The Earth's shadow is approximated as a cylinder; True and eccentric anomalies are calculated for the points at which it is intersected by the orbit. 17
2.10. Screenshot from the 1979 video game "Asteroids": A specially designed vector graphics processor was used for drawing the on-screen objects. 18
2.11. The 1993 video game "Starfox" used a graphics coprocessor called the "Super FX" chip to accelerate the rendering of 3D polygons such as the space ship. . 19
2.12. "Quake" was one of the first PC games to support modern 3D graphics hardware. The overlay on the left side shows the polygons used to construct the scene. 20
2.13. Illustration of the fixed-form GPU pipeline (based on [Kirk and Hwu, 2010], figure 2.1). 21
2.14. Architectural differences between CPU and GPU hardware based on [NVIDIA Corporation, 2015], figure 3. 22
2.15. Screenshot from "The Witcher 3" showing realistic movement of hair and foliage which can be calculated on a GPU. 23
2.16. Screenshot from the upcoming space flight game "Star Citizen" showing asteroids and space debris: Realistic visualization of clouds, lighting and material surface properties rendered in real-time on a modern GPU. 24

2.17. Example of CUDA's thread organization with two-dimensional blocks and grids based on [Kirk and Hwu, 2010], figure 3.13. 26
2.18. Illustration of thread and array layout for the above kernel. 27
2.19. Overview of the different types of memory CUDA offers. Based on [Kirk and Hwu, 2010], figure 3.9. 31
2.20. UML diagram of the Satellite class and the two child classes derived from it. . 34

3.1. Performance of the shader and CUDA code versus the CPU ([Möckel et al., 2011]). Results are combined from two experiments with different population sizes which is why the CUDA curve has more data points than the others. . . 38
3.2. Plugin Context as shown in [Marquardt, 1999]. 41
3.3. The modified plugin design pattern upon which OPI is built. 42
3.4. UML diagram of the OPI interface connecting a host application (DOCTOR) and a propagator (Ikebana). For reasons of clarity, not all classes are shown. . 43
3.5. Simplified UML diagram of *OPI::Population* and the associated data types *OPI::Orbit*, *OPI::ObjectProperties* and *OPI::Vector3*. 44
3.6. Simplified UML depiction of *OPI::Host*, the class that specifies the host interface. 45
3.7. Design pattern for an analytical propagator. 46
3.8. Simplified UML diagram of *OPI::Propagator* and *OPI::PerturbationModule*, both derived from the *OPI::Module* class. 47
3.9. Simplified UML diagram of the *OPI::CudaSupport* class. 53
3.10. Visualization of the real-time spatial partitioning and collision risk assessment in DOCTOR. 55

4.1. UML diagram of Ikebana. 65
4.2. Flowchart of the *runPropagation* function in Ikebana's main class. 66
4.3. Flowchart of Ikebana's luni-solar PerturbationModule. 72
4.4. Flowchart of Ikebana's solar radiation pressure PerturbationModule. 73
4.5. Handling of atmospheric data in FLORA. 74
4.6. Handling of atmospheric data in Ikebana. 75
4.7. Flowchart of Ikebana's Atmospheric PerturbationModule (including functions from AtmosphericData). 76

5.1. The reference population used for propagator validation and analysis, visualized with DOCTOR. 77
5.2. Comparison between Ikebana and FLORA with and without OPI. Very small differences between the two FLORA versions are caused by slightly different rounding of the output data. 82
5.3. Comparison of the zonal harmonics module of Ikebana and FLORA on a LEO orbit. Object number: 13464 83
5.4. Comparison of the zonal harmonics module of Ikebana and FLORA on a GEO orbit. Object number: 28472 83
5.5. Comparison of the third body perturbations module of Ikebana and FLORA on a LEO orbit. Object number: 13464 84
5.6. Comparison of the third body perturbations module of Ikebana and FLORA on a GEO orbit. Object number: 28472 85

5.7. Third body perturbations: Slight deviations in eccentricity caused by the code shown in listing 5.2. Object number: 22963 86
5.8. Solar Radiation Pressure: This LEO orbit comparison shows several of the observed effects such as an oscillating semi major axis and small deviations in inclination and RAAN. Object number: 37452 88
5.9. Solar Radiation Pressure: Small deviations occur in all orbital elements. Object number: 34242 89
5.10. Solar Radiation Pressure: Error handling for negative eccentricities can lead to unpredictable results. Object number: 38098 90
5.11. Atmosphere: Slight deviations in eccentricity, inclination, and RAAN. Object number: 28075 92
5.12. Atmosphere: The semi major axis declines slightly faster in Ikebana. Object number: 33504 93
5.13. Atmosphere: Decline of the semi major axis in high-eccentricity GTO orbits Object number: 25850 94
5.14. Atmosphere: Decline of the semi major axis in high-eccentricity GTO orbits Object number: 36833 95
5.15. Total Results: Histogram showing the error rates between FLORA and Ikebana. 97
5.16. Total Results: Histogram showing the error introduced into FLORA's output when the F10.7 values for solar activity are overestimated by 2%. 98
5.17. Total Results: Histogram showing the error introduced into FLORA's output when the atmospheric density values are overestimated by 3%. 98
5.18. Total Results: Error rates of 1000 random objects. 99
5.19. Total Results: Typical LEO object with slightly faster decline of the semi major axis and no major changes in the other elements. Object number: 28075 100
5.20. Total Results: The faster decline of the semi major axis in the atmospherical model causes an overall slightly increased decay rate. Object number: 39769 . 101
5.21. Total Results: LEO objects with imminent decay show relatively large deviations of semi major axis and eccentricity at the final data point. Object number: 3757 102
5.22. Total Results: Some highly eccentric orbits with low perigees show relatively large deviations of semi major axis and eccentricity. Object number: 22670 . . 103
5.23. Total Results: Without the atmospherical model, the inaccuracies visible in figure 5.22 have largely vanished. Object number: 22670 104
5.24. Total Results: Propagation of GEO orbits is generally free of deviations. Object number: 22963 105
5.25. Total Results: The deviations introduced by the solar radiation pressure module (compare figure 5.9) are small enough to get cancelled out. Object number: 34242 106
5.26. *Vanguard-1*. 107
5.27. Vanguard-1 propagated with FLORA and Ikebana. The dashed line shows the changes caused by setting the drag coefficient to 2.4 which is realistic according to [Bowman, 2002]). 107
5.28. Vanguard-1 TLE data plotted against FLORA and Ikebana results. 108
5.29. Performance in megapropagations per second of FLORA and Ikebana on various platforms. 114
5.30. Relative speedup of Ikebana on various platforms compared to FLORA (i7-3820). 114
5.31. Percentage of the total run time of each perturbation module. 118

5.32. Performance of FLORA and Ikebana with double precision: All platforms but the Tesla suffer significant performance losses. 119
5.33. Relative speedup of Ikebana with double precision compared to FLORA. . . . 120
5.34. Ikebana with double precision: Most objects show no significant deviation from the single precision version (LEO). Object number: 37452 122
5.35. Ikebana with double precision: Most objects show no significant deviation from the single precision version (GEO). Object number: 22963 123
5.36. Ikebana with double precision: In other cases, the results from FLORA are reproduced more closely. Object number: 12952 124
5.37. Ikebana with double precision: The low eccentricity problem of Ikebana can be addressed by using double precision (GTO). Object number: 37847 125

6.1. Ground track projection in DOCTOR. 127
6.2. Simplified UML diagram of DOCTOR. 128
6.3. Flowchart illustrating DOCTOR's initialization and main loop. 129
6.4. The GUI's scripting console allows direct input of Lua functions for controlling animation and setting PropagatorProperties for OPI plugins. 131
6.5. Illustration of the GPGPU propagation from [Möckel et al., 2011]: The orbital data and parameters were provided as textures and used by a vertex shader to calculate the true anomaly. 133
6.6. Ikebana propagating 150,000 objects in DOCTOR at almost 40 frames per second. 136

7.1. UML diagram of the proposed *OPI::PopulationModifier* plugin type. 138
7.2. Example of an *OPI::PopulationModifier* used to add launch traffic to a population. 138

B.1. Error rates of the zonal harmonics module. 171
B.2. Error rates of the lunisolar module. 172
B.3. Error rates of the solar radiation pressure module. 173
B.4. Error rates of the atmospherical module. 174
B.5. Combined error rates of zonal harmonics, lunisolar and solar radiation pressure modules. 175

List of Abbreviations

A2M	Area-to-Mass Ratio
AES	Advanced Encryption Standard
ALU	Arithmetic Logic Unit
API	Application Programming Interface
BOL	Beginning of Life
CPU	Central Processing Unit
CUDA	Compute Unified Device Architecture
DOCTOR	Display of Objects Circulating in Terrestrial Orbits
DRAM	Dynamic Random Access Memory
ECC	Error Correcting Code
EOL	End of Life
EUV	Extreme Ultraviolet
FLORA	Fast Long-term Orbit Analysis
FPS	Frames per Second
FPU	Floating Point Unit
GCC	GNU Compiler Collection
GEO	Geostationary Earth Orbit
GLSL	(Open)GL Shading Language
GNU	GNU's Not Unix
GPGPU	General Purpose GPU Computing
GPU	Graphics Processing Unit
GUI	Graphical User Interface
JD	Julian Date
LMRO	Launch- and Mission-Related Objects
MASTER	Meteoroid And Space Debris Terrestrial Environment Reference
HAMR	High Area to Mass Ratio
MJD	Modified Julian Date
MLI	Multi-Layer Insulation
MP/s	Megapropagations per second

NaK	Natrium-Potassium
NaN	Not a Number
NASA	National Aeronautics and Space Administration
NORAD	North American Aerospace Defence Command
NRLMSISE-00	Naval Research Laboratory Mass Spectrometer and Incoherent Scatter Exosphere 2000
OpenACC	Open Accelerators
OpenCL	Open Compute Library
OpenGL	Open Graphics Library
OpenMP	Open Multi-Processing
OPI	Orbital Propagation Interface
PPU	Physics Processing Unit
RAAN	Right Ascension of the Ascending Node
RAM	Random Access Memory
SDK	Software Development Kit
SDL	Simple DirectMedia Layer
SIMD	Single Instruction, Multiple Data
SPACE	Scientific Parallel Animation and Computation Environment
SRP	Solar Radiation Pressure
SSA	Space Situational Awareness
SSR	Satellite Situation Report
TDP	Thermal Design Power
TLE	Two-Line Elements
UML	Unified Modeling Language
ZUNIEM	Zuschlag Numerical Integration of the Equations of Motion

List of Symbols

Orbital Elements

a	Semi major axis	m
e	Eccentricity	-
i	Inclination	rad
Ω	Right ascension of the ascending node	rad
ω	Argument of perigee	rad
M	Mean anomaly	rad
E	Eccentric anomaly	rad
v	True anomaly	rad
p	Semilatus Rectum ("orbital parameter")	m

Benchmarking

$\check{M}$	Propagator Benchmark Index	$\frac{MP}{s}$
N_P	Number of propagation operations	-
N_{proc}	Number of parallel processors/threads	-
N_{obj}	Number of population objects	-
N_{st}	Number of propagation time steps	-
P_{par}	Parallelizable section of an algorithm	%
S	Speedup factor	-
ΔT	Propagation step size	s
T_P	Propagation time frame	d
t_{st}	Propagation time per object per time step	s
t_{total}	Total propagation time	s

Orbital Physics

A	Cross-sectional area	m^2
A_p	Daily planetary amplitude	-
$\vec{a}$	Aerodynamic acceleration	$\frac{m}{s}$
a_p	Planetary amplitude	-
$\vec{a}_{SR}$	Solar radiation acceleration	$\frac{m}{s}$
B	Ballistic coefficient	$\frac{kg}{m^2}$
C_D	Drag coefficient	-
C_R	Reflectivity coefficient	-
$F_{10.7}$	Radio intensity at 10.7cm wavelength	-
g	Gravitational acceleration	$\frac{m}{s^2}$
h	Scale height	m
m_{sat}	Satellite's mass	kg
p_A	Atmospheric pressure	$\frac{N}{m^2}$
p_{SR}	Solar radiation pressure	$\frac{N}{m^2}$
M	Molar mass	$\frac{kg}{mol}$
R	Gas constant	$\frac{J}{mol \cdot K}$
R_E	Mean radius of Earth	m
$\vec{r}_{sat}$	Radius vector between satellite and Sun	m
S	Shadow function	-
T	Temperature	K
v_{sat}	Satellite's orbital velocity	$\frac{m}{s}$
λ	Longitude	deg
μ	Gravitational parameter of Earth	$\frac{km^3}{s^2}$
ρ	Atmospheric density	$\frac{kg}{m^3}$
ϕ	Latitude	deg

Abstract

Orbital debris is becoming an increasing problem for space flight missions. New satellite launches, explosions, collisions and other events cause a steady rise in the number of objects orbiting the Earth. It is therefore important to determine the future development of the object population, as well as the effectiveness of debris mitigation measures, in long-term simulations. Orbital propagation, the calculation of an object's movement in its orbit, poses a challenge for this research due to the high computation times of the complex perturbation models involved. Sophisticated analytical methods exist that are able to propagate an object in mere milliseconds per time step, including perturbations from atmospheric drag, the Earth's uneven gravitational field, third bodies and solar radiation pressure. However, due to the high population sizes of hundreds of thousands of objects as well as simulation time frames of up to 200 years, these calculations can still take up hours of computation time. To speed up this process, the analytical propagator *Ikebana* is introduced in this thesis. It was programmed to run on graphics processing units (GPUs), hardware designed for massively parallel execution of up to thousands of concurrent threads. This reduces the overall run time for large object populations from hours to minutes. Porting software from a conventional CPU is not a trivial task and involves a number of potential pitfalls and optimization opportunities which are detailed in this work. The propagator is integrated into other applications via a generic, multi-platform interface specifically designed for this task. It allows to develop the propagator separately and integrate it into other tools as a plugin at run time. The interface's architecture serves as a design template for analytical propagation software. It also features automated mechanisms that facilitate the development of GPU-based propagators which compute the motion of large object numbers in parallel. As an exemplary application that uses *Ikebana* as a plugin, the space debris visualization tool *DOCTOR* is introduced. Its real-time propagation requirements and use of the graphics processor were the sparks that started this research. With the parallelized propagator integrated via the generic interface, the software is able to fluently animate populations of hundreds of thousands of objects while taking into account all relevant perturbations.

The title image shows satellites and space debris objects larger than one centimetre, visualized with DOCTOR and propagated one month into the future with *Ikebana*. One object is highlighted to show its current orbit and position (yellow) and its original orbit (white).

The project on which this thesis is based has been funded by the German Federal Ministry for Economic Affairs and Energy under the grant number **50 RA 1306**. The responsibility for the contents of this work lies with the author.

Zusammenfassung [German Abstract]

Weltraumschrott wird für die Raumfahrt ein zunehmendes Problem. Da sich die Anzahl der Objekte im Weltall durch neu gestartete Satelliten, Kollisionen und Explosionen kontinuierlich erhöht, ist es wichtig, mittels Langzeitsimulationen die zukünftige Entwicklung der Objektpopulation und die Wirksamkeit von Müllvermeidungsmaßnahmen abzuschätzen. Die Bahnpropagation, d.h. die Berechnung der Bewegung der Objekte auf ihren Umlaufbahnen, verursacht dabei erheblichen Rechenaufwand. Mit analytischen Verfahren kann die Bahnbewegung eines Zeitschritts inklusive Störeinflüsse durch Atmosphäre, Erdgravitation, Drittkörper und solaren Strahlungsdruck in Sekundenbruchteilen berechnet werden; bei der Betrachtung ganzer Populationen von mehreren Hunderttausend Objekten über Zeiträume von bis zu 200 Jahren fallen dennoch Rechenzeiten von mehreren Stunden an. Um diesem Problem Herr zu werden, wird in dieser Arbeit der analytische Bahnpropagator *Ikebana* vorgestellt, der für die Ausführung auf Grafikprozessoren (GPUs) konzipiert wurde. Bei diesen Prozessoren handelt es sich um hochparallele Recheneinheiten, die in der Lage sind, viele Hundert Objekte gleichzeitig zu berechnen und somit die Gesamtrechenzeit für große Populationen erheblich zu reduzieren. Bei der Portierung der Software von einer gewöhnlichen CPU sind einige hardwarespezifische Hürden zu beachten, die ebenfalls detailliert beschrieben werden. Der Propagator wird über eine eigens dafür entwickelte, generische Schnittstelle in Anwendungsprogramme eingebunden. Diese erlaubt es, den Propagator getrennt zu entwickeln und als Plugin bereitzustellen, das zur Laufzeit in beliebige Simulationstools integriert werden kann. Die Architektur der Schnittstelle dient dabei als Designvorlage für analytische Propagationssoftware; ebenso verfügt sie über Automatismen, die die Entwicklung GPU-basierter Propagatoren erleichtert. Als beispielhaftes Anwendungsprogramm wird das Visualisierungstool *DOCTOR* vorgestellt, das die Weltraummüllpopulation darstellt. Sein Anspruch an den Propagator, große Populationen in Echtzeit zu berechnen sowie die Verwendung von Grafikhardware bilden den Ursprung dieser Arbeit. Durch die Einbindung des parallelisierten Propagators über die generische Schnittstelle wird das Programm in die Lage versetzt, Populationen von mehreren Hunderttausend Objekten unter Berücksichtigung aller relevanten Störeinflüsse flüssig zu animieren.

Das Titelbild zeigt Satelliten und Weltraumschrott der Zentimeterpopulation, visualisiert mit DOCTOR unter Verwendung von Ikebana als Propagator. Die Population wurde einen Monat in die Zukunft propagiert; ein Objekt ist hervorgehoben, dessen aktuelle und ursprüngliche Umlaufbahn in gelb, bzw. weiß dargestellt sind.

Das dieser Arbeit zu Grunde liegende Forschungsvorhaben wurde mit Mitteln des Bundesministeriums für Bildung und Forschung unter dem Förderkennzeichen **50 RA 1306** gefördert. Die Verantwortung für den Inhalt dieser Veröffentlichung liegt beim Autor.

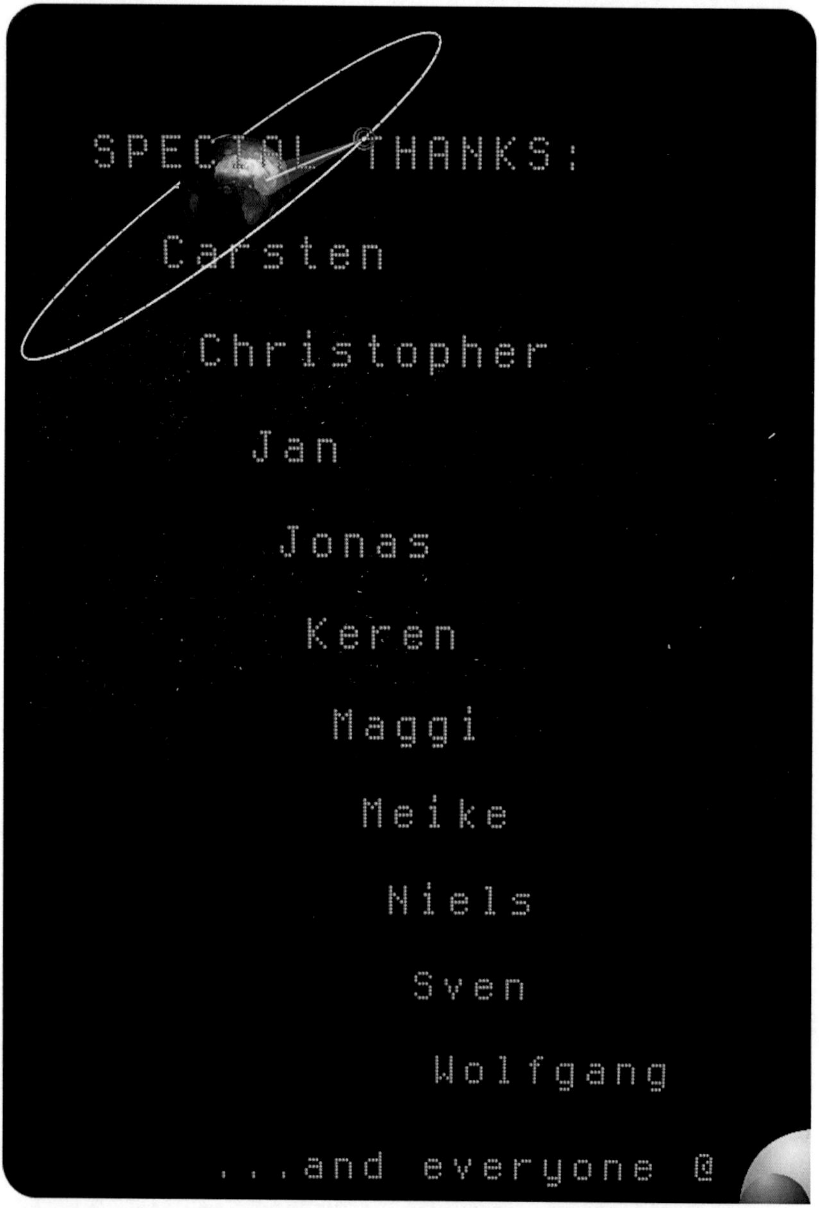

SPECIAL THANKS:

Carsten

Christopher

Jan

Jonas

Keren

Maggi

Meike

Niels

Sven

Wolfgang

...and everyone @

Screenshot of DOCTOR showing the wonderful people who made this possible. Thank you so much!

1 Introduction

> People write research papers about, "Here's a really tricky algorithm to do that", but it doesn't work in all cases. People who are synthesists and think of those complex algorithms, they really pooh-pooh that. They don't like to hear that because they want it to fall to cleverness rather than raw power; but the way things have consistently, undeniably fallen over the years is to raw power.
>
> John D. Carmack

John Carmack became known to the space community when he founded Armadillo Aerospace in 2000. His fame, however, stems from a different profession: He is recognized as one of the world's leading innovators in the field of real-time 3D graphics engines for video games ([MIT Technology Review, 2002]). The statement above was made in 2004 when he was interviewed about the "Doom III" graphics engine and his plans for following projects ([Kent, 2004]). It comments on the fact that complex computational problems can often only be solved by more powerful hardware. While some can be addressed efficiently with sophisticated algorithms, these often make simplifying assumptions that will only work under a narrowly defined set of conditions; a proper solution can only be realized once the technology becomes powerful enough to process the additional amount of data. However, he goes on to recognize that "cleverness" is still required to harness this power in the most effective way so the amount of simultaneously processable problems can be increased. This is especially true in the field of digital entertainment where tough competition forces developers to get the most out of existing hardware in order to deliver the best product.

Obviously, this does not just apply to entertainment applications. Video games use computers to create a simulation of the real world; even though the content is usually fictional, people expect things to look and feel in a certain way based on their real-life experience, such as gravity, physical conditions of depicted objects, shadows and reflections. The more detailed the model, the higher the immersive effect. Scientific applications work in a very similar way: A real-world entity is described by a mathematical model and simulated with a computer in order to study it. The more accurate the simulation, the better the results can be applied to real world problems. However, increased detail in the model almost always comes at the cost of higher computational effort. For this reason, most scientific disciplines have greatly profited from the advances that computer science has made over the last decades. With the ever-increasing computational power of modern PCs, simulations can be executed at higher speed and with more detail, data from measurement campaigns can be processed faster and more information can be stored and organized.

In many cases however, simply relying on faster computers to improve the speed of an existing algorithm is not sufficient. This has become especially apparent early in this millenium when a paradigm shift took place in the way that computer hardware was designed. During the 1980's and 1990's, hardware manufacturers focused mainly on increasing the clock speed of their processors to improve their power. Recently, a physical limit has been reached where it is no longer feasible to further improve the clock rate due to the excessive heat generation that it involves ([Kirk and Hwu, 2010]). Instead, the main area of optimization now lies in increasing parallelism (i.e. adding more "cores" to a processor). This forces software developers to adapt their algorithms to this new paradigm: Most applications that existed up to then performed their operations in a strictly sequential order. In order to benefit from additional cores, such applications must be redesigned to execute some of their instructions in parallel.

3D graphics applications are particularly suitable for parallelization since the vertices and pixels from which a scene is drawn can be processed independently from each other. For this reason, graphics processors are optimized for massively parallel execution. With the advent of general purpose programming interfaces for these devices, they can be used to simulate physical models not just for entertainment but also for science. In space research specifically, orbital propagation of large object populations can greatly benefit from this. The physical models used in those algorithms are similar in nature to the three-dimensional graphics applications: Both make extensive use of trigonometric functions and vector operations in 3D space. In addition, since the individual objects of the population do not usually interact with each other, propagation can easily be executed in parallel.

1.1. Space Debris

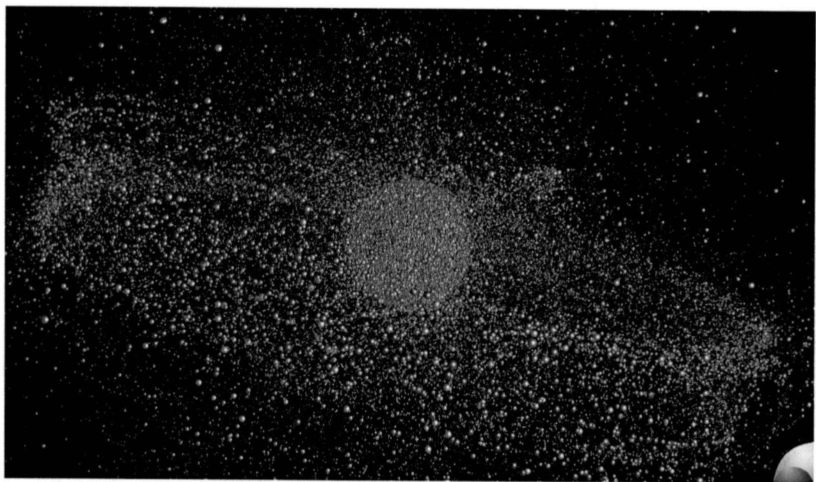

Figure 1.1.: Illustration of the >1cm object population in Earth orbit as of 2009.

Almost all space flight missions produce debris. At the end of their operational lifetime, most satellites in higher orbits remain in space because the energy required to return them to Earth would be much too high. As of 2009, only about 25 per cent of the approximately 3200 catalogued satellites were operational. Depending on the orbit characteristics, some would take hundreds or even thousands of years to slow down and burn up in the atmosphere; others, specifically objects in the Geostationary Earth Orbit (GEO), have virtually no chance of returning. Those objects are usually moved into higher, so-called *graveyard orbits* at the end of their lifetime to make room for new operational spacecraft. But spent satellites are not the only source of debris. Some of the upper stages of the rockets that were used to place them into orbit do not return to Earth and remain in space. Earlier spacecraft often used explosive charges for operations like removing lens caps from telescopes. While this practice has largely been ceased for reasons of debris mitigation, some of the debris generated by it can still be observed today. Apart from these *launch- and mission-related objects (LMRO)*, secondary sources of mostly smaller debris particles include flakes of paint caused by surface degradation, pieces of multi-layer insulation (MLI) foil delaminating from satellites and rocket bodies, droplets of leaked sodium-potassium (NaK) used as a coolant, and dust and slag from solid rocket motor firings. While larger objects can be observed by radars and telescopes, most of them are too small to be detected. ESA's *MASTER (Meteoroid and Space Debris Terrestrial Environment Reference)* application uses physical models that simulate the sources of these undetectable particles in combination with observation data to generate its object database. The data is statistical in nature so it cannot be used to assess actual object positions. It is however possible to calculate collision probabilities and possible long-term developments based on it. Overall, the documentation ([Flegel et al., 2011]) lists over 220,000 representative[1] objects in the size regime of one millimetre as well as approximately 150,000 larger than 5 mm, ca. 116,000 objects larger than 1 cm (figure 1.1), ca. 28,000 objects larger than 10 cm and over 5,000 objects larger than one metre. Numbers for object sizes down to $1\mu m$ are in the millions. All objects are potentially hazardous and may pose a risk to operational spacecraft. Clouds of small particles can damage solar panels and cause the satellite to malfunction; but especially the larger ones can be extremely dangerous. Due to the high velocities at which they are traveling an impact with a piece of debris as small as 1cm in diameter can potentially fragment the satellite causing more debris. Figure 1.2 shows a simulation of a collision between spacecraft: In the event, more than 5,000 fragments were generated. At first they are distributed along the satellites' original orbits; perturbation forces cause them to spread over time and cover wider areas, potentially crossing into other spacecraft's orbits.

In most orbital regions, collision probabilities are still low. But new launches, explosions and collisions continuously increase the population. In 1978, [Kessler and Cour-Palais, 1978] described a cascading effect that would occur in orbital regions with a high population density: If a critical number of objects was reached, collisions among them would cause the population to increase even though no further objects were added by space flight missions. This is commonly known as the *Kessler Syndrome*; the data from MASTER 2009 (figure 1.3) shows that the altitudes around 800 kilometre are at the highest risk.

Space debris has become a global problem that all major space agencies worldwide aim to solve in collaboration. Recent studies conducted at the Institute of Space Systems ([Möckel et al., 2013], [Möckel et al., 2015]) show that countermeasures such as post-mission disposal and active removal of high-risk objects can help to reduce the impact of debris on the near-Earth environment. This thesis aims to make a small contribution by introducing an orbital propagator that is capable of handling the huge object numbers involved in this research.

[1] *Representative* refers to the practice of subsituting several small objects with similar properties for a single one.

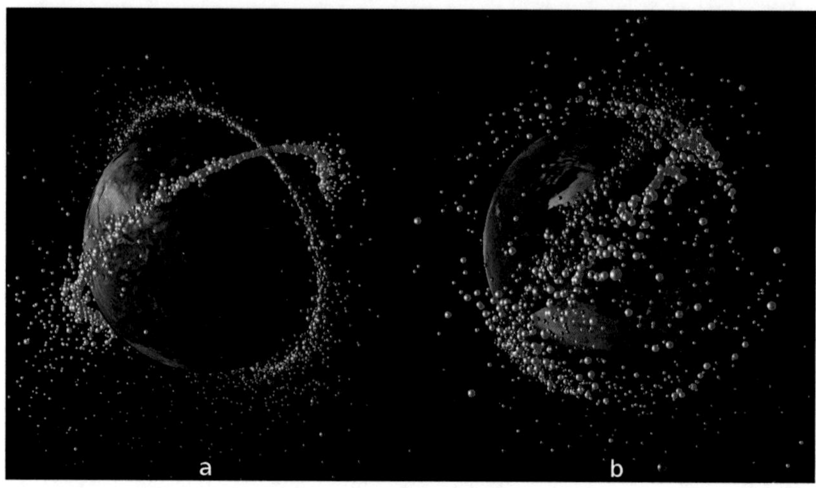

Figure 1.2.: Simulation of a collision between two satellites (a) minutes after the event and (b) six months later, showing the distribution of the generated fragments.

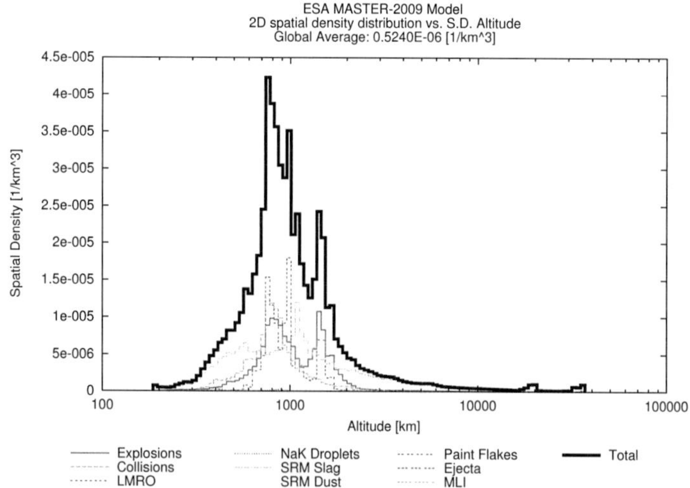

Figure 1.3.: A spatial density plot from MASTER 2009 showing various sources of space debris. The region around the 800 kilometre altitude band shows the highest number of objects.

1.2. Scope of Work

1.2.1. Computational Models for Orbital Propagation

Orbital propagation is a mathematical process by which a satellite's orbit around a celestial body and its position on that orbit are determined for a given time. Like all computational models, no approach is capable of accurately describing reality in every detail. Increasing the complexity of the model always results in higher computation time. For any given problem, a trade-off must be found that is both sufficiently accurate and sufficiently fast. In computer graphics, for example, a circle can be approximated by a polygon with evenly distributed edges; the more edges, the closer the resemblance. Drawing a perfect circle would require an infinite number of edges and, subsequently, infinite computational time. However, from a certain point onwards, the polygon and the perfect circle will be visually indistinguishable; any additional time spent on drawing more edges will be wasted.

For orbital propagation, as with many other applications, several computational models exist which can be broadly divided into three distinctive categories: Numerical, analytical and semi-analytical. [CCSDS, 2010] provides a good overview of these three categories: Numerical propagation is performed by direct numerical integration of the differential equations of motion describing the forces that influence the object's position and velocity. While generally very accurate, numerical propagation is also the most computationally complex method. The underlying equations describe the change in position and velocity for a given time step based on the previous output; obtaining a solution for a specific time requires both an initial position and velocity as well as a continuous iterative process. Just like in the above example, the solution can be further approximated by adding more steps, i.e. more computational effort, until the desired accuracy is reached. As illustrated in figure 1.4, the maximum achievable accuracy is limited by the scope of the underlying model.

Analytical propagation is based on mathematical formulas that describe an object's position as a function of its orbit and the desired time. They are often derived from real-world observations, for example, by performing a regression analysis on large sets of measurement data. In contrast to numerical propagation, a solution can be obtained directly for any given point in time. However, analytical models describe the underlying physical effects in a much simpler manner. Therefore, in general, analytical propagation is computationally faster but less accurate. Since no approximation process takes place, accuracy improvements can only be achieved by revising the model itself.

Semi-analytical propagation is a combination of the two methods that can be used if accurate orbit determination over long time periods is required. Only the long-term perturbation effects are taken into account which are described analytically, but with more accurate models as in the case of full analytical propagation. These are then integrated numerically. This allows for relatively accurate estimation of the object's orbit but also makes it possible to use larger time steps.

1.2.2. Use Cases

Orbital propagation forms the basis of many different topics in space research. Long-term predictions of the space debris environment are done by propagating a catalogue of objects into the future while analyzing possible events such as breakups, collisions, new launches and decays. But even to build such an object catalogue, object propagation is required: During

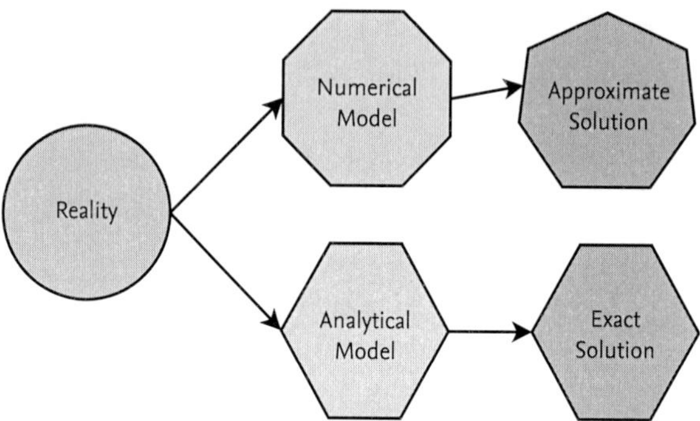

Figure 1.4.: Illustration of the practical difference between analytical and numerical models.

tracking campaigns with radars or telescopes, a spotted object's trajectory has to be calculated in order to predict the location where it will be visible next.

The methods of orbital propagation, as described previously, differ in two key properties: Speed and accuracy. Depending on the requirements given by the propagator's use case, a numerical or analytical method is chosen. This section presents four exemplary use cases and their demands in the properties mentioned above. This work covers analytical propagation which mainly applies to use cases 3 and 4. These two are used as examples throughout the following chapters; the intent of the propagator introduced in chapter 4 is to meet the requirements of both.

Use Case 1: Reentry A spacecraft is about to reenter into the atmosphere. To determine potential locations of debris hitting the Earth's surface, the object is constantly monitored. Predictions are made so that in case of a potential breakup over a populated area a warning can be issued to the public. Many parameters necessary for these predictions, such as the object's tumbling rate and the atmospheric density at its current location are burdened with high uncertainties. However, given the object's high speed and altitude, small variations in the predicted impact location can make a huge difference regarding the object's criticality. To introduce as little error as possible into the calculation, the propagator's accuracy requirement should be regarded as high. Since usually only one object has to be accounted for and location updates are made available only a few times per day, run times of a few minutes or even hours are acceptable. Speed requirements are therefore low. A highly accurate numerical propagator is best suited for this task.

Use Case 2: Tracking and Cataloguing During a tracking campaign, objects in space are followed using an appropriate tracking device such as a radar, telescope or laser. The size of the window in which an object is visible largely depends on the technique used but in any case, only a fraction of the orbit can be captured. To uniquely identify and catalogue an object, multiple detections have to be made and linked to the same object. Usually, this process involves estimating the full orbit from the visible fraction and searching a preexisting catalogue for possible candidates immediately after a detection is made. The tracking device can then be adjusted to scan the position where the object is thought to be visible next. The computational demand for the propagator in this use case is high both in terms of speed and accuracy.

Propagation has to be fast enough to allow the identification of a possible candidate before the object passes over the tracking site again; depending on the technique used, additional time is required to adjust the tracking device accordingly. During this time, large portions of the catalogue may have to be considered if the number of possible candidates is high. The level of accuracy required depends on the size of the detection window but can generally be considered to be high, albeit over much shorter time periods than in a long-term population analysis. In addition to computational power, object tracking is extremely demanding with regard to the bandwidth of networks, memory and storage devices as large amounts of data have to be streamed to the algorithms in near real-time.

Use Case 3: Long-term Analysis For long-term evaluation of the space debris environment, an entire object population is propagated over large time periods of up to 200 years. Depending on the scenario requirements, population sizes of 50,000 to 200,000 objects are considered. Once each time step, typically ranging between one day and three months, satellite launches, collisions and debris removal measures are simulated to estimate the increase in object numbers over time. A Monte-Carlo approach is used to account for uncertainties in the input data. This quickly results in billions of propagation steps that have to be carried out, often occupying computers for hours or even days. It is obvious that this use case requires a propagator that can process large object populations at high speeds which only an analytical method can provide. Since long-term scenarios evaluate the space debris population statistically, the knowledge of the objects' exact position is not required, contrary to the reentry use case. More important are the properties and positions of the orbits themselves from which information such as the number of objects that travel through certain orbit regimes can be deviated. This subsequently enables further analyses such as collision risk assessment. The accuracy of the propagator has to be high enough to correctly reflect the overall development of the population including orbit changes and decay rates of objects. Long-term analyses like this have been carried out in several projects such as [Flegel et al., 2010], [Möckel et al., 2013] and [Möckel et al., 2015].

Use Case 4: Visualization For demonstration purposes, real-time 3D visualization programs can be used to show the movement of single objects or large object populations around the Earth. In order to provide a smooth animation, the desired number of objects has to be propagated at least 30 times per second. While this is easily performed on a regular CPU for one or even a few thousand objects with an analytical propagator, animating a population of hundreds of thousands of objects requires extremely high propagation speeds that are currently available only on massively parallel hardware architectures such as graphics processors. Using analytical equations for this task also has the advantage that they do not depend on values calculated for previous time steps. Arbitrary jumps in time are possible so users can watch the population at different points in time. In that case, the propagator must still be accurate enough to realistically reflect changes in the population such as decays that occur during the skipped time frame without taking longer than a few seconds. Since the objects' positions are merely calculated to show their position on the screen, the visualization case has a low demand for the propagator's accuracy. Especially among large populations small inaccuracies that occur in the position of each object will be undetectable by the user and will not change the overall visual impression. A software tool suitable for this use case will be introduced in chapter 6 of this thesis.

Others A sophisticated software tool that covers two or more of the above use cases might require different approaches to orbital propagation running at the same time, or some method to arbitrarily switch between algorithms. For example, if a tool is created with the purpose of following a specific object's course over time while taking into account possible collision threats from space debris it might be necessary to simultaneously propagate a single object with high accuracy and a large population of other objects with lower accuracy.

Secondly, although not a use case in itself, another important issue should be considered. When a propagator or a component such as a physical model is revised it is necessary to update and validate the implementation. Validation is usually carried out by comparing propagation results against actual orbital data or, if no such information is available, another propagator that has already been validated. This process can be extremely tedious and error-prone, especially if the reference uses a completely different data format.

	Requirement	
	Speed	**Accuracy**
Reentry campaigns	low	high
Tracking campaigns	high	high
Long-term analysis	high	medium
Visualization	high	low

Table 1.1.: Summary of the different propagator use cases and requirements.

1.3. Outline

In chapter 2, the basics of orbital physics are presented as well as the physical models behind the different perturbation forces that act on objects in Earth orbits. The basics of graphics hardware as well as the programming models and techniques specific to these devices are outlined. Chapter 3 describes a software architecture that can be used for analytical propagators with special emphasis on parallel computing. It presents an implementation of this architecture as the software framework *OPI* that can be used for the creation of orbital propagators and applications that require their output to function. A practical example is documented in chapter 4: This tool, called *Ikebana*, is an analytical propagator based on existing software that was redesigned to make use of the software framework as well as parallel computing techniques. Chapter 5 gives a comparison of the enhanced software and its original form and analyzes the differences in terms of execution speed and accuracy. Finally, chapter 6 describes the visualization software *DOCTOR* which animates objects in Earth orbits. Since it relies on fast orbital propagation of large object populations it serves as a reference use case for both *OPI* and *Ikebana*.

2 State of the Art

2.1. Orbital Physics and Propagation

As described by Johannes Kepler in the early 17th century, the movement of a satellite orbiting around a central body assumes the form of an ellipse with the central body in one of its focal points (figure 2.1). The point on the ellipse that is closest to the central body is called *periapsis*, the point that is farthest away is called *apoapsis*. For Earth-centric orbits these points are often called *perigee* and *apogee*, derived from the Greek word for Earth, *gaia*. The size and shape of the ellipse is defined by two parameters. The first is the *semi major axis* (a) which is defined as the length between the center and either periapsis or apoapsis. The second is the *eccentricity* (ϵ) which describes the ratio of the distances "center to focal point" and "periapsis to focal point". It defines the "flatness" of the ellipse: An eccentricity of zero means that the focal point lies exactly in the center and the ellipse is circular. The eccentricity increases the further the focal point moves towards the perimeter; the ellipse becomes a parabola at an eccentricity of one and a hyperbola for larger values. In reality, all closed orbits are elliptical and perfect circular orbits do not exist. Most analytical equations assume that

$$0 < \epsilon < 1 \tag{2.1}$$

and are undefined for eccentricities of zero. The distance from the focal point to the perimeter perpendicular to the semi major axis is called the *semilatus rectum* or simply the *orbital parameter*, p:

$$p = a(1 - \epsilon^2) \tag{2.2}$$

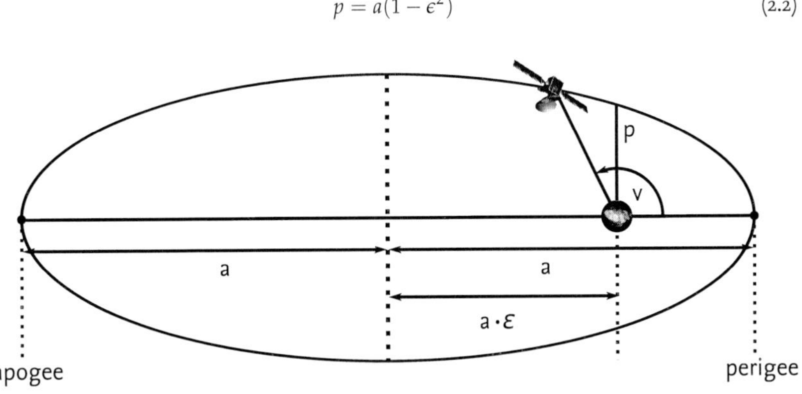

Figure 2.1.: Parameters of an orbital ellipse (based on [Wiedemann, 2014]).

To describe the position of the orbit in space, three additional parameters are defined that are shown in figure 2.2: The *inclination* (i) is the angle between the Earth's equatorial plane and the orbital plane. The *right ascension of the ascending node* (Ω, also abbreviated *RAAN*) is the angle between the vernal point and the point at which the orbit crosses the equatorial plane in ascending direction. The vernal point is defined at the vector from the Earth's center to the Sun's center at the time of spring equinox at which the Sun crosses the equatorial plane. This occurs at different times around March 20th each year. Finally, the *argument of perigee* (ω) describes the angle between the intersection of the orbit's ascending node with the equatorial plane and the orbit's perigee. An additional parameter called the *true anomaly* (v) defines the position of the satellite on the orbit as the angle from its perigee. Instead of the true anomaly, sometimes the *eccentric anomaly* (E) or the *mean anomaly* (M) are used. The eccentric anomaly is illustrated in figure 2.3; it is defined as the angle to the satellite's equivalent position on an auxiliary circle around the ellipse. The mean anomaly can be derived from Kepler's second law which states that a satellite passes over equal areas of the orbital plane in equal time (figure 2.4). On an elliptical orbit this means that the speed is highest at the perigee and lowest at the apogee. The mean anomaly describes the angle to the position that the satellite would assume on a circular orbit with an equal semi major axis after an equal amount of time; the closer the eccentricity of an orbit gets near zero, the smaller the difference between M and v. While the true and eccentric anomalies can only be calculated iteratively, the mean anomaly can be expressed as

$$M = \sqrt{\frac{\mu}{a^3}} \cdot t \tag{2.3}$$

where μ is the Earth's gravitational parameter and t is the time in seconds since the last crossing of the perigee.

2.1.1. Perturbation Forces

In an ideal environment, a once established orbit would not change without explicit action taken by the spacecraft. In reality, however, several external forces act on the satellite which cause its orbit to deviate from its original form and position. They stem from various sources and affect different orbital parameters with different intensities. The key element in orbital propagation is to simulate these forces as accurately as possible. Many different models and approaches exist that offer a tradeoff between precision and speed if run time is an issue. Perturbation forces can be divided into *periodic* and *secular* perturbations. Short periodic effects cause orbital elements to "jitter" around their mean values; for example, the varying gravitational pull from the Earth has such an effect on the eccentricity with frequencies in the hour range. Other effects are long periodic - for example, the gravitation from Sun and Moon causes the inclination of GEO objects to oscillate over a period of around 53 years. Secular perturbations are those that invoke a continuous deviation on the orbit, such as the atmospheric drag decreasing the semi major axis of the satellite.

The following sections describe the most influencial perturbation forces and how they can be dertermined analytically.

2.1.1.1. Zonal Harmonics

For Keplerian orbits it is assumed that the central planetary body is a perfect sphere with an equally distributed gravitational potential. In reality, the Earth is slightly flattened at the poles, i.e. the equator is slightly elliptical and a little longer than the longitudinal diameter. Likewise, the Earth's mass is not uniformly distributed resulting, for example, in a higher gravitational potential in the presence of large mountain ranges. The real gravitational potential affecting the satellite must therefore be expressed as a function of its position. In his

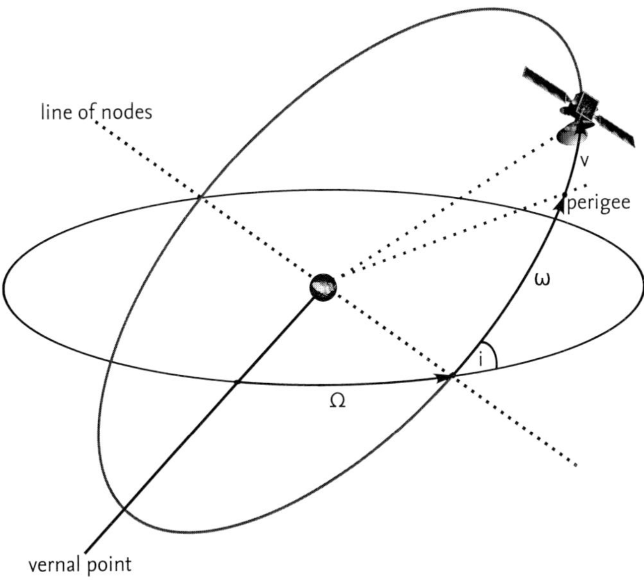

Figure 2.2.: Parameters describing the position of the orbit and the satellite (based on [Wiedemann, 2014]).

compendium, [Vallado, 2007] gives the equation for the gravitational potential, called U, as a function of the satellite's latitude (ϕ), longitude (λ) and radius from the Earth's center (r):

$$U = \frac{\mu}{r}[1 + \sum_{l=2}^{\infty} \sum_{m=0}^{l} (\frac{R_E}{r})^l \cdot P_{l,m} \sin(\phi) \cdot (C_{l,m} \cos(m\lambda) + S_{l,m} \sin(m\lambda))] \tag{2.4}$$

$P_{l,m}$ are Legendre polynomials into which the satellite's position factors. $S_{l,m}$ and $C_{l,m}$ are coefficients that describe the Earth's deviation from a perfect spherical form. For $m = 0$ they define zonal harmonics, i.e. bands of latitude (figure 2.5 left); $l = m$ applies to sectoral harmonics (bands of longitude, figure 2.5 middle) and $l \neq m$ to tesseral harmonics ("checker-board", figure 2.5 right). $C_{l,0}$ is often used as a negative and substituted:

$$-C_{l,0} = J_l \tag{2.5}$$

J_2 includes the Earth's equatorial bulge which has the strongest influence by far. More J_l-terms can be added to account for further deviations and subsequently improve the model's accuracy. Satellite-based measurement campaigns conducted to evaluate the Earth's gravitational potential allow to derive constant values for these terms. [Vallado, 2007] cites analytic equations for the affected orbital elements that rely solely on these constants as well as the satellite's current orbit. Zonal harmonics exert long-periodic perturbations on semi major axis and eccentricity, secular perturbations on RAAN and argument of perigee (figure 2.6) as well as short-periodic perturbations on all orbital elements.

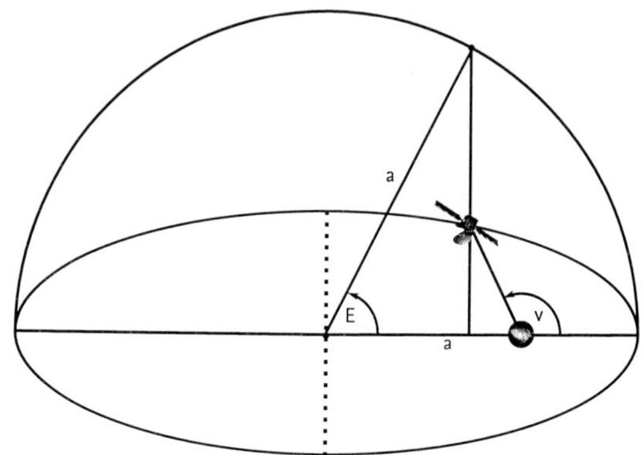

Figure 2.3.: Illustration of the eccentric anomaly (based on [Wiedemann, 2014]).

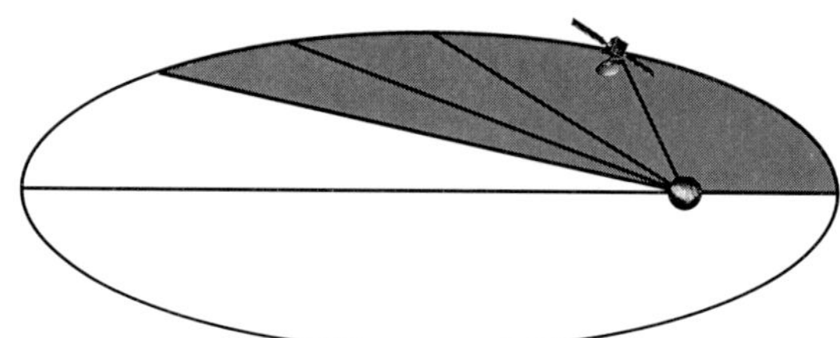

Figure 2.4.: Illustration of Kepler's second law based on [Vallado, 2007]. The grey areas have the same size.

2.1.1.2. Atmospheric Drag

A satellite moving through the Earth's atmosphere experiences a drag force exerted by particles hitting its front-facing area. This causes the semi major axis and eccentricity to decrease lowering the orbit's apogee continuously (figure 2.7). Near-circular orbits also experience periodic perturbations in inclination, RAAN and argument of perigee.
The aerodynamic drag affecting a satellite can be expressed as an acceleration vector ([Vallado, 2007]):

$$\vec{a} = -\frac{1}{2}\frac{\rho v_{sat}^2}{B} \tag{2.6}$$

v_{sat} is the satellite's velocity and can be derived from its orbital parameters. Since the value is relative to the atmosphere which moves with the Earth, the absolute velocity can be approximated by subtracting the Earth's mean motion. B is called the *ballistic coefficient* and is defined as the satellite's mass m_{sat} divided by the product of its drag coefficient C_D and the cross-sectional area facing in the direction of the satellite's velocity A:

$$B = \frac{m_{sat}}{C_D A} \tag{2.7}$$

Figure 2.5.: Zonal, sectoral and tesseral harmonics

Since the exact orientation of a satellite is usually difficult to determine, A is often approximated as the mean area of the object in a randomly tumbling state. For high-precision orbital propagation, information from the satellite's sensors can be used if available. Similar to the area, the drag coefficient depends on the satellite's shape and configuration. In many cases, an approximate value of 2.2 can be used. B is usually regarded as a constant when a randomly tumbling state is assumed and loss of mass due to spent fuel is neglected.

ρ is the atmospheric density at the satellite's position and is usually the parameter that is most difficult to be determined. It is closely linked to the atmosphere's temperature (T) and pressure (p_A) at a given position; the respective equations are given by [King-Hele, 1987]:

$$\frac{p_A}{\rho} = \frac{RT}{M} \tag{2.8}$$

with R being the gas constant ($8.31 J K^{-1} mol^{-1}$) and M being the molar mass of the gas. The pressure decreases with increasing height which is expressed by the *hydrostatic equation*:

$$\frac{dp_A}{dy} = -\rho g \tag{2.9}$$

with g being the gravitational acceleration. Eliminating ρ between these equations yields

$$\frac{dp_A}{p_A} = -\frac{Mg}{RT} dy \tag{2.10}$$

RT/Mg is called the *scale height*:

$$h = \frac{RT}{Mg} \tag{2.11}$$

It can be assumed as constant and used to approximate the amount of pressure loss within discrete height ranges.

The atmospheric density is primarily influenced by variations in the Earth's magnetic field as well as the Sun's extreme ultraviolet (EUV) radiation. EUV rays heat the upper layers of the

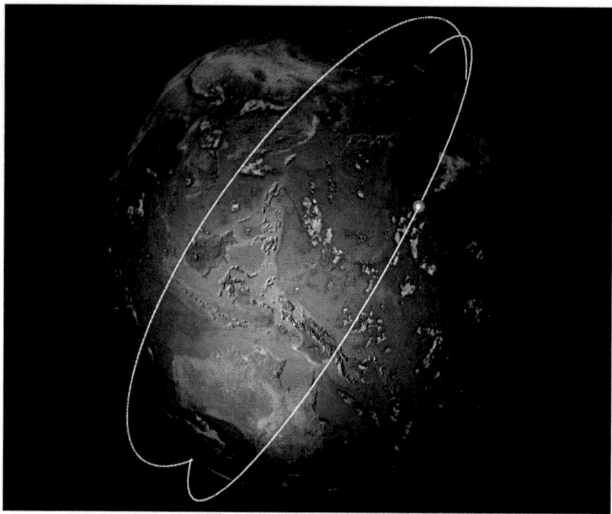

Figure 2.6.: Visualization of the RAAN change caused by zonal harmonics perturbations.

atmosphere and therefore increase the density and, in turn, cause satellites to experience a stronger drag. Although the EUV radiation itself is difficult to determine, it correlates with the solar emission's radio intensity at a wavelength of 10.7 cm ($F_{10.7}$) which can be measured and used as an indicator for solar activity. In a similar manner, particles from the Earth's magnetic field cause heat by ionization effects in the upper atmosphere which affects the density ([Vallado, 2007]). Geomagnetic activity is measured around the world and recorded every three hours as the *planetary amplitude*, a_p; the daily average over the eight values is called the *daily planetary amplitude*, A_p. While measurements are possible for magnetic field and the solar activity, predictions for future dates are very difficult.

For analytical propagation, several models have been developed that approximate atmospheric density based on these measurements as well as other necessary parameters such as current time and the satellite's position. A popular example is the *Naval Research Laboratory Mass Spectrometer and Incoherent Scatter Exosphere 2000 (NRLMSISE-00)* model described in [Picone et al., 2002]. It is based on the earlier *MSIS-86* model which was extended to include the exosphere (*MSISE*) and augmented with data from satellite missions of the US Naval Research Laboratory. It is an empirical model based on an extensive database of information gathered from measurements of several spacecraft and radar campaigns. Analytical equations were derived from analyzing this data and published with an accompanying open-source Fortran application. It takes as input a date and time, a position expressed as geodetic latitude, longitude and altitude, current and averaged $F_{10.7}$ solar flux as well as the A_p index. From these, the model calculates the atmospheric density at those coordinates. It also outputs individual masses of the various elements that make up the atmosphere as well as the temperature; this information can be inserted into equation 2.11 as M and T, respectively, to calculate the scale height.

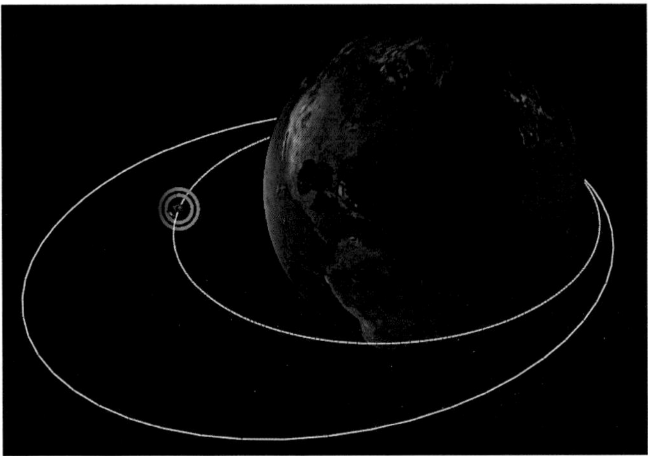

Figure 2.7.: Visualization of the apogee decrease caused by atmospheric drag.

With approximate values for scale height and atmospheric density, the analytic equations given by [King-Hele, 1987] can be used to calculate the changes in orbital parameters. Different models are given based on the orbit's eccentricity. For near-circular orbits with an eccentricity lower than 0.2 all orbital parameters are affected by the atmosphere. For highly-eccentric orbits only semi major axis and eccentricity need to be taken into account.

2.1.1.3. Third Body Perturbations

Kepler's equations describe a two-body problem, i.e. all objects other than the central body and the satellite are disregarded. In reality, other massive bodies in the vicinity extent a gravitational pull on the satellite drawing it out of its orbit. For objects around the Earth, the most influencial third bodies are the Sun due to its enormous mass, and the Moon due to its relatively short distance. Perturbations from Sun and Moon cause long-periodic changes in eccentricity, inclination, RAAN and argument of perigee, as well as secular changes to the last two and the mean anomaly. The semi major axis remains largely unaffected. A notable example of such an effect is the long-periodic disturbance of the GEO orbit mentioned before: Due to the way the ground tracks of the third bodies are inclined relative to Earth, their gravitational pull causes the inclination of GEO objects to oscillate by approximately ± 15 degrees over a period of 53 years (figure 2.8); this forces operators of telecommunications satellites to regularly adjust their positions.

[Vallado, 2007] presents analytical equations that give the rate of change in degrees per day for all affected orbital elements as a function of the third body's position. The position is given as a vector (A, B, C) depending on four parameters of the third body: The equatorial inclination i_3 and right ascension of the ascending node Ω_3 relative to Earth, its argument of mean longitude u_3 and its distance from the Earth's center r_3. Equations for these parameters for both Sun and Moon are given by [Flegel, 2007] as a function of the Julian date. As a fifth parameter, the equations require the gravitational parameter of the disturbing body μ_3 which, together with the distance, influences the strength of the gravitational pull.

Figure 2.8.: Simulation of third body effects causing an inclination buildup in GEO orbits.

2.1.1.4. Solar Radiation Pressure

Every satellite exposed to sunlight experiences perturbations which are caused by impacting particles from solar rays. The pressure which they exert is assumed constant and is given by [Vallado, 2007]:

$$p_{SR} = 4.57 \cdot 10^{-6} \frac{N}{m^2} \tag{2.12}$$

The actual acceleration force which this pressure applies on the satellite largely depends on its properties. The surface area that is exposed to the Sun influences the amount of rays that hit the satellite, its mass determines the impact: The larger the exposed area and the smaller the mass, the bigger the perturbation. Like for the atmospheric perturbations, the exact area is difficult to be determined so a mean value is used; loss of mass over time is neglected. Since these two values depend on each other in both cases, a combined value called the *area-to-mass (A2M) ratio*, given in square metres per kilogram, or its inverse is often used for a spacecraft. A high area-to-mass ratio causes a larger impact of solar radiation pressure. Another important property is the reflectivity coefficient, C_R, which acts as a multiplier to the force based on the amount of rays that are reflected or absorbed. Values range from zere to two: A value of zero means that no force is transmitted (i.e. the satellite is translucent). 1.0 implies that all rays are absorbed and the force is fully transmitted (i.e. a black surface). If all rays are reflected by the surface twice the force is applied which is expressed by a C_R value of 2.0. Since the actual coefficient is very difficult to determine a mean value of 1.3 is often used. With these variables, [Vallado, 2007] gives the acceleration as

$$\vec{a}_{SR} = -p_{SR} C_R \frac{A}{m_{sat}} \cdot \frac{\vec{r}_{sat}}{|\vec{r}_{sat}|} \tag{2.13}$$

$\vec{r}_{sat}$ is the radius vector from the Sun which is also the direction in which the force is applied. The source also contains formulas based on a work by [Cook, 1962] expressing the perturbation caused by the force in RSW (radial, transverse, normal) components which can be used to determine the changes in the orbital elements. Solar radiation pressure affects all orbital elements although the effects are very small for most objects and are therefore often concealed by larger perturbations such as atmospheric drag. A notable exception is MLI foil which has a very high area-to-mass ratio (see [Flegel, 2013]).

These equations apply when the satellite is exposed to full sunlight. However, during its travel on its orbit, most satellites cross the Earth's shadow at some point. A different set of equations can be used for this case which require the true and eccentric anomalies of the shadow entry and exit, respectively, as additional inputs. Determining these requires knowledge of the satellite's orbit and current position, as well as the position and geometry of the Earth's shadow. [Escobal, 1965] gives a simplified model for this which assumes a cylindrical shadow; umbra and penumbra effects are neglected as well as the Earth's oblateness and movement in its orbit (figure 2.9). With these simplifications, a shadow function can be employed:

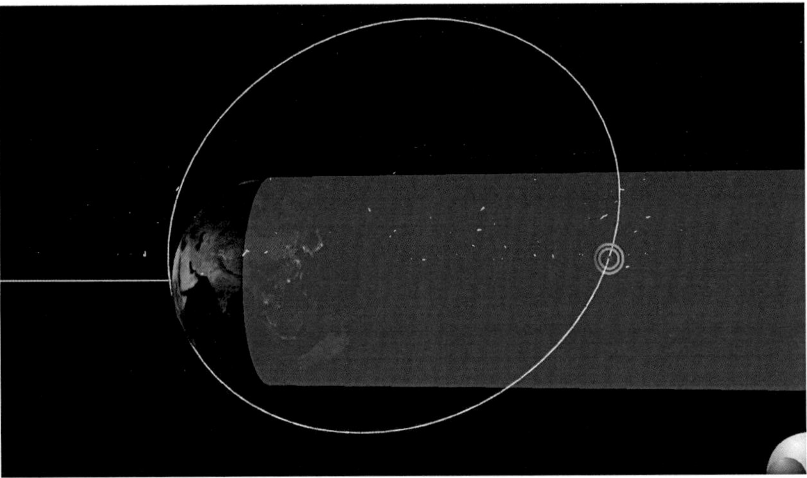

Figure 2.9.: Illustration of the simplified shadow model. The Earth's shadow is approximated as a cylinder; True and eccentric anomalies are calculated for the points at which it is intersected by the orbit.

$$S \equiv {R_E}^2(1 + e\cos v)^2 + p^2(\beta\cos v + \xi\sin v)^2 - p^2 \tag{2.14}$$

with β and ξ being the dot products of the Sun's position with the unit vectors pointing to the current argument of perifocus and true anomaly, respectively. This equation can be transformed to become

$$S^* = A_0\cos^4 v + A_1\cos^3 v + A_2\cos^2 v + A_3\cos v + A_4 \tag{2.15}$$

with coefficients depending on R_E, p, β and ξ. The roots of this function can be found using a quartic equation solver and checked against equation 2.14 for validity: If S is evaluated to zero and $\beta \cos v + \xi \sin v < 0$, v gives the true anomaly of a shadow entry or exit. Which of these applies can be determined through comparison with surrounding values.

2.2. GPU Computing

2.2.1. A Short History of Graphics Processors

Figure 2.10.: Screenshot from the 1979 video game "Asteroids": A specially designed vector graphics processor was used for drawing the on-screen objects.

In the field of computer graphics, one of the significant driving factors behind the technical advances of the last three decades has been the video game industry (for which space travel, as for virtually all branches of fiction and entertainment, has always been a popular topic). Early video games such as Atari's "Asteroids" (figure 2.10), although quite simple in nature, already used sophisticated hardware to produce images on the screen. The game was released as an arcade machine in 1979 and used a digital vector generator that was designed by Atari specifically for a small number of games. It was built from basic circuitry such as digital-analog-converters, latches and clock generators and can translate a binary representation of a 2-dimensional vector into screen coordinates; [Margolin, 2001] describes the hardware in great detail.

Since the late 1970's, video games have constantly increased in popularity and have largely moved from arcade machines with dedicated hardware tailored specifically to a certain game to multi-purpose home computers and game consoles. Over the years, video games became more and more demanding from the hardware in order to improve the quality of graphics and simulation as well as the quantity of content. The willingness of people to upgrade their devices every few years to be able to play the latest games opened up a mass market for computer manufacturers and stimulated research on graphics hardware. Games that were distributed on hardware cartridges sometimes featured dedicated graphics coprocessors such as the "Super FX" chip used by Nintendo in the early 1990's to complement the console's builtin hardware. Compared to the early vector processor from the "Asteroids" game, it was able to draw vector images in three dimensions allowing for more realistic objects. It also moved from simple mesh representations to polygons with colored faces (figure 2.11).

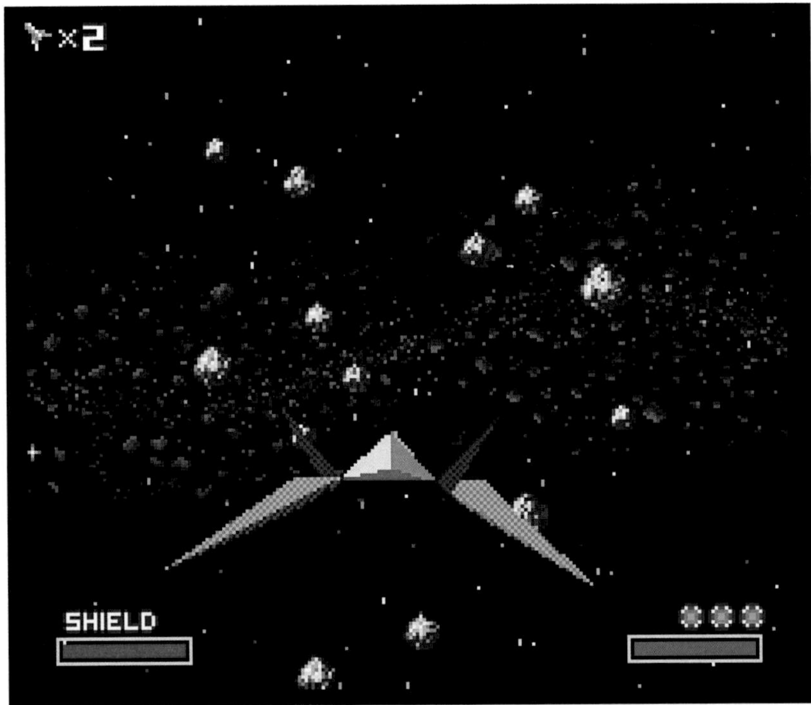

Figure 2.11.: The 1993 video game "Starfox" used a graphics coprocessor called the "Super FX" chip to accelerate the rendering of 3D polygons such as the space ship.

The mid-1990's saw the rise of 3D computer graphics not just in video games but also in other forms of entertainment. In the PC sector, dedicated 3D graphics processing units (GPUs) that were previously only found in large workstations used to create pre-rendered graphics for movies and television shows could be made affordable and used on PC video cards with real-time capability. In 1996, the PC game "Quake" (figure 2.12) was released; its graphics engine developed by John Carmack is believed to have played a major role in triggering this development: Contrary to similar applications released at the time the game's graphics were generated solely from fully textured, three-dimensional polygons that were rendered in real time. The initial version relied on the CPU to perform the necessary computations. In later versions support for 3D graphics hardware was added that relieved the CPU of the complex vector operations and allowed for higher framerates and smooth interpolated textures. On the software side, application programming interfaces (APIs) such as the *Open Graphics Library* (OpenGL) and Microsoft's *Direct3D* were created that provided an abstraction layer from the hardware and enabled software developers to write applications that would run on all supported GPUs.

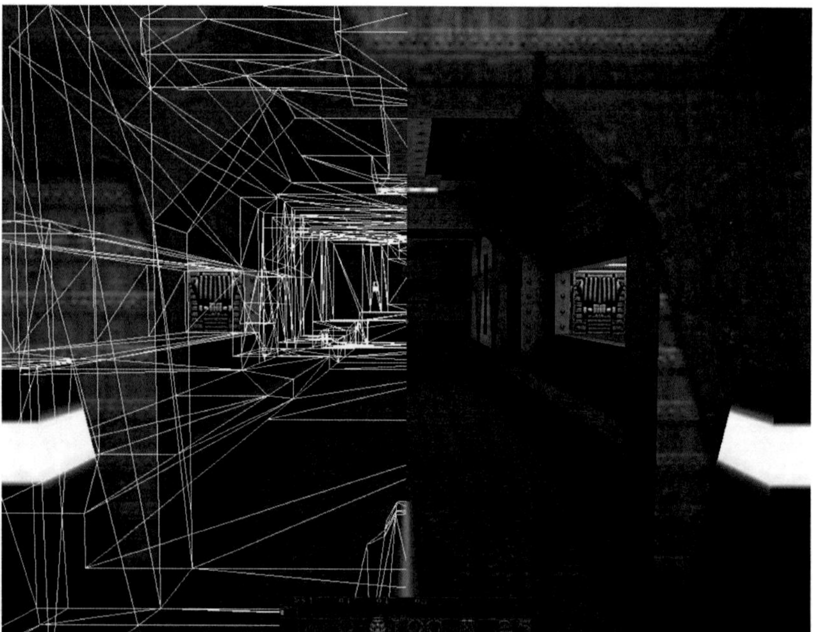

Figure 2.12.: "Quake" was one of the first PC games to support modern 3D graphics hardware. The overlay on the left side shows the polygons used to construct the scene.

The programming model for these early GPUs was based on a pipeline structure that processed incoming commands in a fixed order. This is described in [Kirk and Hwu, 2010] and shown in figure 2.13. The CPU delivers geometry information that is broken down into triangles and submitted as a list of vertices that form the corners of those triangles. Image data that is used for texturing is uploaded to the GPU's texture memory. The pipeline can be separated into two main stages. The "vertex stage" first transforms the geometry data into a hardware-native format (*vertex control*). It then assigns values such as colors, normals and texture coordinates to each vertex (*shading, transform and lighting*) and prepares them for rasterization (*triangle setup*). The "pixel stage" performs the task of translating the three-dimensional geometry information into a two-dimensional screen projection. It first calculates which pixel of the resulting image is influenced by which triangle (*raster*). Each of these pixels is assigned a color based on the per-vertex information as well as relevant texture images and lighting configuration (*shader*). The *raster operation* stage performs final calculations for effects such as antialiasing and transparency before the final image is written to the frame buffer and displayed on the screen.

The vertex stage and the pixel stage are responsible for two aspects that greatly influence the quality of the resulting image. The more vertices the hardware can process, the finer the geometry that can be constructed. In a similar manner, a higher pixel density improves the visual quality of the final image. Since both the per-vertex and the rasterization operations can be performed for each vertex/pixel individually, the GPUs were optimized for massively parallel execution of those calculations. In both stages, the algorithm working on each vertex or pixel is the same, just with different input data. This is called the *Single Instruction, Multiple Data (SIMD)* principle[1].

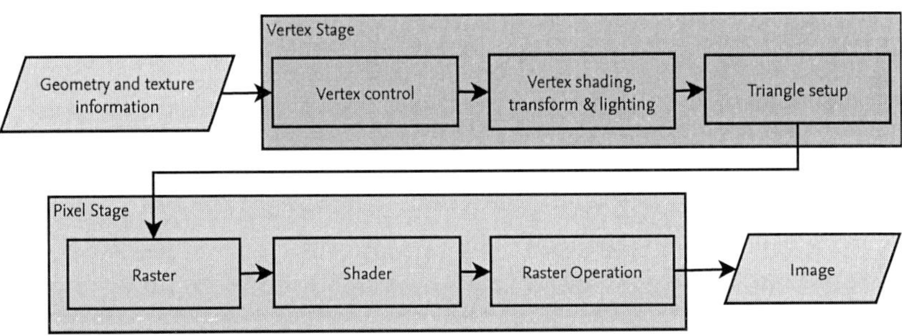

Figure 2.13.: Illustration of the fixed-form GPU pipeline (based on [Kirk and Hwu, 2010], figure 2.1).

With the newly opened mass market for 3D graphics processors and still increasing customer demand for better graphics quality the performance and capabilities of these devices were improved continuously. Apart from increasing clock speed, parallelism, bandwidth and texture memory, a major new development took place in the early 2000's when the fixed-form pipeline which offered little flexibility was made programmable. Since then, APIs such as OpenGL allow software developers to directly influence the pipeline stages with small programs called *shaders* that are executed directly on the GPU. In OpenGL, the C-like *OpenGL Shading Language (GLSL)* is defined for this purpose. Shader programs consist of two com-

[1]Also referred to as *SIMT (Single Instruction, Multiple Thread)* in the context of actual hardware implementations

ponents: *Vertex shaders* interfere with the transform and lighting stage; they are run on each vertex of the scene and can be used to modify each vertex' position, texture coordinates, normals, per-vertex color and more. *Fragment shaders*[2] are run on each pixel of the resulting image; information from the vertex stage is available and can be used to influence the color. Shader programs offer a large amount of flexibility and can be used to create advanced graphical effects such as reflections on water surfaces ([Truelsen, 2007]), sub-surface scattering for realistic depiction of human tissue ([d'Eon et al., 2007]) and ambient occlusion for realistic shadows ([Szirmay-Kalos et al., 2010]). Shader programs are usually loaded from files at runtime and compiled and uploaded to the GPU by the graphics API.

Because of the distinctive hardware, programming a GPU requires an alternative approach compared to a CPU. Figure 2.14 illustrates the architectural differences between the two. Both designs consist of the same components: The dynamic random access memory (DRAM) from which data and program code are fetched; the cache memory which is located directly on the processor and used to store data which is required immediately for fast access; the control section responsible for instruction management and scheduling, and a number of arithmetic logic units (ALUs) which execute the actual instructions. The main difference lies in the arrangement of these components. Since the CPU is targeted at providing a wide range of functions its instruction set and flow control mechanisms are much more complex than that of the GPU which is tailored towards a very specific application. The CPU's cache and control section are therefore much larger. The GPU is optimized at performing a large number of relatively simple instructions in a short time which means that many more transistors are devoted to performing arithmetic operations. The ALUs are organized in clusters with their own memory and control sections which allows for more efficient management of large streams of input data. Usually, they are arranged into cores and paired with dedicated floating point units (FPUs) to allow fast, parallel processing of floating point numbers ([NVIDIA Corporation, 2009]).

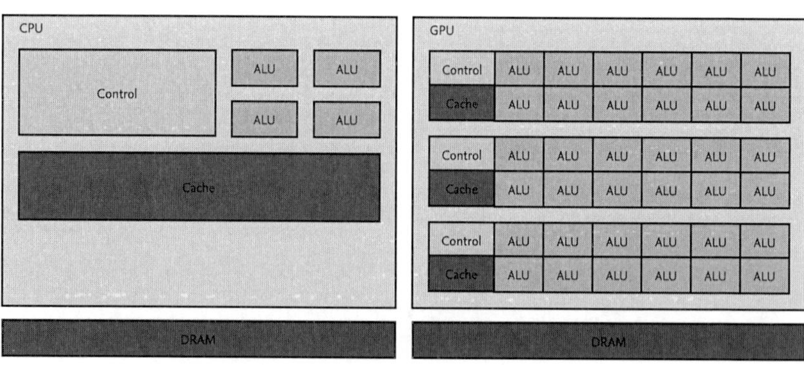

Figure 2.14.: Architectural differences between CPU and GPU hardware based on [NVIDIA Corporation, 2015], figure 3.

[2] Also called *pixel shaders* in the Direct3D equivalent of the GLSL.

Other than providing high-quality graphics, many video games also rely on a realistic depiction of the environment. Basic features such as collision detection and movement, as well as advanced techniques such as force models, fluid dynamics and other particle effects such as smoke, fire and explosions rely on the computation of physical models. While such calculations do not have to be highly accurate they have to be executed at very high speeds - in addition to the rendering - to provide a smooth frame rate. In the early 2000's companies started developing physics processing units (PPUs), such as the *Ageia Physx*, to relieve the CPU of such calculations. While no actual hardware specifications were published, [Blachford, 2006] approximates the design based on a US patent that was filed for this purpose. He found that the hardware architecture was very similar to that of graphics processors; this makes sense because especially particle effects rely heavily on SIMD vector processing, just like 3D graphics. With the advent of programmable graphics pipeline on GPUs, these processors became able to calculate physical effects implemented as shader programs; existing physics APIs were modified to use the GPU instead of, or in addition to, dedicated PPUs. For example, the game "The Witcher 3" can use the GPU for fast calculation of particle effects to simulate realistic movement of foliage and strands of hair in the wind (figure 2.15).

Figure 2.15.: Screenshot from "The Witcher 3" showing realistic movement of hair and foliage which can be calculated on a GPU.

Today, most personal computers are equipped with dedicated, programmable graphics processing units that compute sophisticated visual and physical effects in real-time, producing high-resolution images created from millions of triangles and spanning millions of pixels at speeds of 30 to 60 frames per second; dedicated physics processors have largely vanished from the market. Over the last 35 years the visualization of space in video games has changed from simple vectorized objects on a black background to highly realistic depictions of space objects, clouds, planetary surfaces, lights and shadows (figure 2.16).

Figure 2.16.: Screenshot from the upcoming space flight game "Star Citizen" showing asteroids and space debris: Realistic visualization of clouds, lighting and material surface properties rendered in real-time on a modern GPU.

2.2.2. General Purpose GPU Computing

In the mid 2000's the new programmable architecture of the modern GPUs caught the attention of scientists who sought after more efficient ways to solve their complex mathematical problems. For calculations that fit into the SIMD principle, GPUs were much better suited than regular CPUs. More precisely, [Owens et al., 2008] state that applications are suitable to be run on a GPU if:

1. "The computational requirements are large", i.e. the algorithm has to be run on thousands or even millions of samples,

2. "Parallelism is substantial", i.e. there are no direct dependencies between the samples, and

3. "Throughput is more important than latency", i.e. the order in which individual results are output does not matter.

However, as the GPU was designed to produce images from three-dimensional geometry, the APIs did not provide a way to upload or download arbitrary data into the GPU's memory. Uploads were limited to mainly vertex arrays and texture images. Downloads were only possible for texture images or screen sections that were generated on the GPU; since those images stayed in GPU memory for the majority of applications, the associated API functions were relatively slow. For programmers, this meant that they had to arrange their algorithms in a way that they could be represented as graphical problems and solved using vertex and fragment shaders. For this approach, the term *GPGPU (General Purpose GPU) computing* was coined. As an example, [Harrison and Waldron, 2007] describe a method of using the GPU as a coprocessor for the AES encryption algorithm. The plaintext is arranged as a two-dimensional texture on which a shader program performs the encryption. The resulting ciphertext is again stored in texture memory. Implementations like these, while practically useful, require a lot of extra effort and produce code that is difficult to read. To solve this problem, graphics processor manufacturers developed APIs targeted at running massively parallel general purpose applications on GPUs, such as the *Open Compute Library (OpenCL)* and the *Compute Unified Device Architecture (CUDA)*. The latter has been used in the context of this work and is described in the following section. Chapter 6 gives a practical example of the differences between CUDA and GPGPU techniques which are used in the visualization as a fallback option.

2.2.3. CUDA

2.2.3.1. Overview

CUDA is an application programming interface developed by NVIDIA Corporation specifically for their line of graphics processors. It was designed to provide programmers with a method of using the GPU for general purpose computing without the hassle of having to transform their algorithms into graphics problems first. CUDA provides a programming language that is similar to that used for vertex and fragment shaders but it is not limited to graphics applications. It allows direct memory transfers of arbitrary data types between the CPU and the GPU memory in both directions. Just as the underlying hardware, CUDA is optimized for massively parallel SIMD applications. Conceptually, the programming model that CUDA provides is very closely linked to this design; contrary to regular modern programming languages which are largely abstracted from the underlying hardware, optimization of CUDA programs requires a profound understanding of the processor's architecture. It also means that currently existing applications cannot simply be executed on a parallel hardware but parts of the software need to be redesigned. While this is true for other parallel processors such as multicore CPUs, as [Sutter and Larus, 2005] point out, this becomes even more apparent on GPUs because of the larger architectural differences.

2.2.3.2. Build Process

To develop CUDA programs the CUDA Software Development Kit (SDK) is required. It includes the libraries and hardware drivers necessary to execute applications on the supported hardware, the CUDA compiler *nvcc*, debugging tools such as the *Visual Profiler* as well as code samples and documentation. CUDA applications cannot work on the GPU alone; they rely on a portion of the code that is run on the CPU to manage the execution of the GPU part. In this context, the CPU along with the system memory and periphery is called the *host*, the graphics card containing the GPU and its dedicated memory is called the *device*. CUDA functions are embedded into an ordinary programming language, usually C or C++. For compiling, *nvcc* parses the source and separates it into host and device code. The latter is compiled directly into a device-executable CUDA binary (*cubin*), the former is forwarded to a standard compiler such as *gcc*. The resulting object files are linked into a binary for the respective operating system with the device code embedded into it.

2.2.3.3. Kernels and Threads

When developing a CUDA application, everything that can be run in parallel is implemented into the embedded CUDA functions. These functions, called *kernels*, are designed to work according to the SIMD principle: On the device, multiple instances of a kernel (called *threads*)

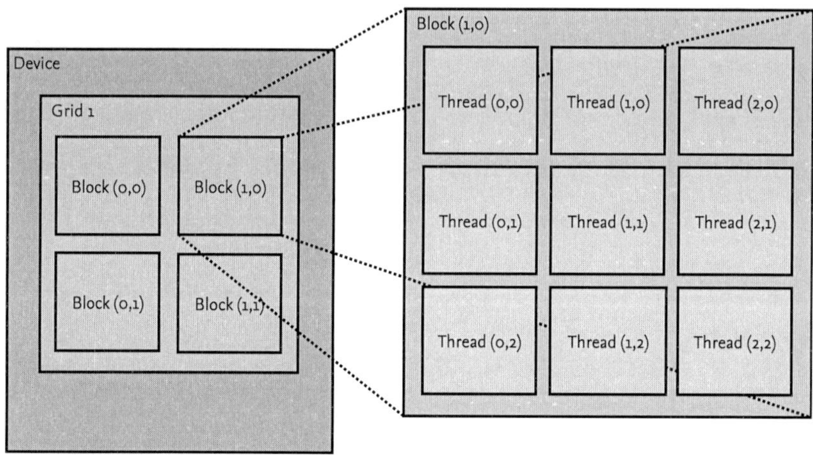

Figure 2.17.: Example of CUDA's thread organization with two-dimensional blocks and grids based on [Kirk and Hwu, 2010], figure 3.13.

are run in parallel with each thread working on one set of input data. The data is initialized on the host, typically in the form of one-, two-, or three-dimensional arrays. The support for up to three dimensions stems from the hardware optimization for graphics programming which mainly operates on multidimensional data such as 2D texture images and 3D Cartesian coordinates. In the same manner, threads need to be arranged into a one- to three-dimensional layout; this is illustrated in figure 2.17 and explained in [Kirk and Hwu, 2010]: When a kernel is executed, all threads that are generated from it are grouped in a *grid*. Each grid consists of multiple blocks which contain the threads. The number and layout of threads in each block and blocks in each grid is configured upon kernel execution. The maximum block and grid sizes are limited by hardware capabilities[3]. Both grid and block layouts can be one-, two-, or three-dimensional; in the above example, two-dimensional layouts are shown for both although individual configuration is possible. CUDA provides several predefined variables which tell each thread its position inside the layout: *gridDim* and *blockDim* contain the respective sizes of grids and blocks in each dimension; *blockIdx* and *threadIdx* contain the threads' block and thread indices. In the above example, *gridDim* would be (2,2,1), *blockDim* would be (3,3,1); indices of the blocks and threads are shown in the figure with the z-dimension omitted.

The grid and block layouts are chosen based on two main factors. The first is the nature of the problem and the format of the input data. In the most simple case, the kernels work on a one-dimensional array. If the threads are organized in a single block with a one-dimensional thread layout, each thread can use its own index as a pointer into the input array. The second factor is memory and run time considerations which will be discussed in more detail in the following section. Listing 2.1 shows a very simple example of a CUDA kernel. It takes as input two one-dimensional integer arrays a and b, adds them and stores the result in array

[3] For the hardware used in the context of this work, the exact numbers can be found in appendix A.

a. For this example to work, the threads need to be laid out by the executing host code in the fashion stated above; each thread then reads its individual index on the x-dimension from the variable *threadIdx.x* and use it as an index into the input arrays. Unless constrained by hardware limits or execution configuration, all elements of the arrays are added in parallel, as illustrated in figure 2.18.

Kernel functions are identified by one of three keywords *__global__*, *__host__* or *__device__*. Global functions run on the device but can only be executed by the host. They serve as entry points for the parallel part of the program. Device functions run on the device and can only be executed by a global function or another device function. These are used as subroutines for more complex kernels which are usually split into one global and several device functions. Host functions are called from and executed on the host; they are used to provide compatibility for machines that have no CUDA hardware available. In this case, the CUDA driver emulates a device on the CPU and executes the host function in this manner. The *__host__* and *__device__* identifiers can be combined; in this case, the compiler creates two versions of the function: One that is run on the device and one that is run in emulation mode automatically in case no suitable hardware is found.

Listing 2.1: A simple CUDA kernel that adds two integer arrays in parallel.

```
// A simple CUDA kernel that adds the values of the two integer arrays a and b
// and stores the result in array a. Every thread of this kernel works on one
// element of each array corresponding to its own index.
__global__ void add(int *a, int *b)
{
  a[threadIdx.x] += b[threadIdx.x];
}
```

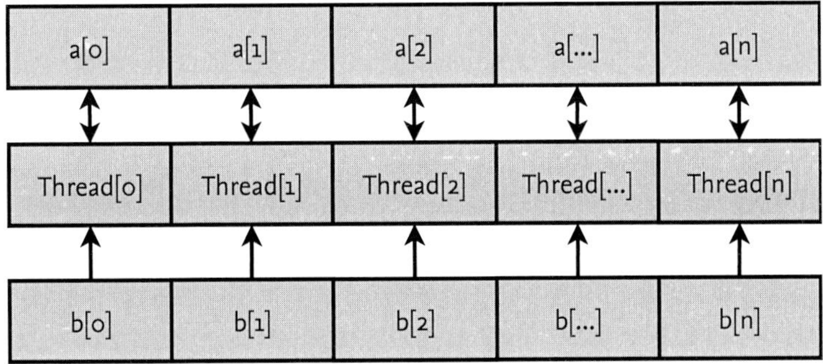

Figure 2.18.: Illustration of thread and array layout for the above kernel.

2.2.3.4. Hosts and Memory Management

Since the GPU has its own memory data needs to be copied between host and device prior to the kernel execution and afterwards to download the results of the computation. The CUDA library provides several host functions for this. The most basic are *cudaMalloc* and *cudaMemcpy*; similar to their C counterparts *malloc* and *memcpy* they can be used to allocate memory on the device and copy data to and from the allocated memory regions. This is shown in listing 2.2 which contains the host code required to run the above kernel: Two integer arrays a and b are created and filled with data. Two additional variables are created that serve as pointers to these arrays on the device. Using *cudaMalloc*, device memory is allocated and the device pointers are initialized; in the two following *cudaMemcpy* calls they are used to copy the arrays to those locations. Next, the grid and block sizes are configured using the *dim3* data type provided by CUDA for this purpose. Since the data is a one-dimensional array with 32 elements, a single grid with a one-dimensional block of 32 threads is chosen. As explained above, this makes it possible to directly use each thread's x-coordinate as an index into the arrays a and b. In the next step, the kernel is called; the grid and block sizes are passed to global kernel functions enclosed in the <<< and >>> identifiers. The kernel is now executed on the device with all 32 instances running in parallel. To pause the host program until all threads have finished their calculations, the function *cudaDeviceSynchronize* can be called. Finally, the results that the kernel stored in array a are copied back to the host and printed to the screen.

Listing 2.2: A simple host program that executes the kernel from the above example.

```
#include <stdio.h>
#include "cuda.h"

// Main function written in C
int main(int argc, char** argv)
{
  // Set length of input array and its size in bytes
  const int length = 32;
  const int bytesize = length*sizeof(int);

  // Initialize input arrays and fill them with data
  int a[length];
  int b[length];

  for (int i=0; i<length; i++) {
    a[i] = i;
    b[i] = i;
  }

  // Create device memory pointers for the input data
  int* dev_ptr_a;
  int* dev_ptr_b;

  // Allocate device memory for arrays a and b
  cudaMalloc((void**)&dev_ptr_a, bytesize);
  cudaMalloc((void**)&dev_ptr_b, bytesize);

  // Copy arrays a and b to the device memory
  cudaMemcpy(dev_ptr_a, a, bytesize, cudaMemcpyHostToDevice);
  cudaMemcpy(dev_ptr_b, b, bytesize, cudaMemcpyHostToDevice);

  // Set block size and grid size for one-dimensional kernel execution:
  // One grid holds one block, each block contains a number of threads
  // corresponding to the size of the input array.
  dim3 dimBlock(length, 1);
  dim3 dimGrid(1, 1);

  // Call the kernel with the given block/grid sizes and input
  add<<<dimGrid, dimBlock>>>(dev_ptr_a, dev_ptr_b);

  // Wait until all threads have finished
  cudaDeviceSynchronize();

  // Results are stored in array a - copy it back to the host
  cudaMemcpy(a, dev_ptr_a, bytesize, cudaMemcpyDeviceToHost);

  // Free memory on the device
  cudaFree(dev_ptr_a);
  cudaFree(dev_ptr_b);

  // Print the results and return.
  for (int i=0; i<length; i++) printf("%d ", a[i]); printf("\n");
  return 0;
}
```

As of version 6, CUDA supports a concept called *Unified Memory* which obviates the need for manual data transfers. Listing 2.3 shows the same host code as the previous example but with the use of unified memory. Instead of using *cudaMalloc* to allocate device memory, the function *cudaMallocManaged* is called to allocate memory on both the host and the device. The same pointers can now be used to fill the arrays with data on the host, add them on the device and print them after a successful kernel execution. CUDA automatically handles the required memory transfers in the background whenever access to this data is requested on either host or device.

Listing 2.3: The same host program using the managed memory available in CUDA 6.

```
#include <stdio.h>
#include "cuda.h"

// Main function written in C
int main(int argc, char** argv)
{
  // Create input arrays
  int *a;
  int *b;

  // Set length of input array and its size in bytes
  const int length = 32;
  const int bytesize = length*sizeof(int);

  // Allocate managed memory for arrays a and b
  cudaMallocManaged((void**)&a, bytesize);
  cudaMallocManaged((void**)&b, bytesize);

  // Fill the arrays with data on the CPU
  for (int i=0; i<length; i++) {
    a[i] = i;
    b[i] = i;
  }

  // Set block size and grid size for one-dimensional kernel execution:
  // One grid holds one block, each block contains a number of threads
  // corresponding to the size of the input array.
  dim3 dimBlock(length, 1);
  dim3 dimGrid(1, 1);

  // Call the kernel with the given block/grid sizes and input
  add<<<dimGrid, dimBlock>>>(a, b);

  // Wait until all threads have finished
  cudaDeviceSynchronize();

  // Print the results on the CPU
  for (int i=0; i<length; i++) printf("%d ", a[i]); printf("\n");

  // Free memory and return
  cudaFree(&a);
  cudaFree(&b);

  return 0;
}
```

CUDA hardware offers different types of memory which differ in size, access speed and scope ([Kirk and Hwu, 2010]). This is illustrated in figure 2.19. Memory transfers from the host go to the *global memory* by default which consists of the graphics card's DRAM. It is by far the largest, but also the slowest memory; typical sizes range from one to four gigabytes on current hardware[4]. All threads have read and write access to it, so large arrays of input and output data are usually stored there. The host also has access to *constant memory* which can be used to store values that do not change over the application's life time. All threads have

[4]The exact sizes of the different memory regions for the hardware used in this work can be found in appendix A.

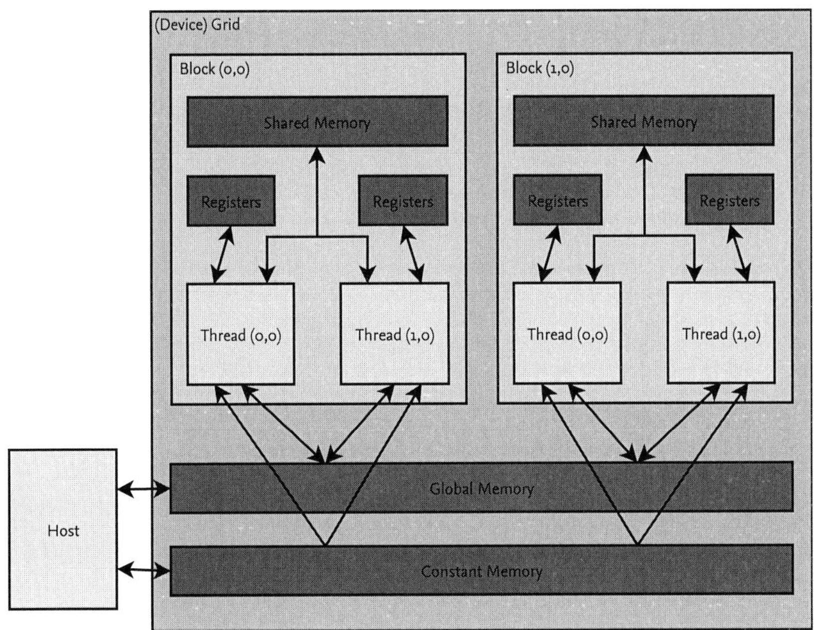

Figure 2.19.: Overview of the different types of memory CUDA offers. Based on [Kirk and Hwu, 2010], figure 3.9.

fast read-only access to it but its size is limited to typically 64 kilobytes on current hardware. It is still useful to store fixed data such as small lookup tables. Adding the *__constant__* identifier to the variable declaration will cause the data to be stored in constant memory. The other types of memory are used internally by the kernels with no direct access from the host. *Shared memory* is scoped on a block level, i.e. all threads of a block share the same instance of a shared variable. It is typically limited to under 100 kilobytes but very fast and highly optimized for parallel access. It can be used to allow communication between the individual threads of a block. Variables can be placed in shared memory by declaring it with the *__shared__* identifier. Registers are the fastest type of memory; local thread variables are usually assigned to it automatically by the compiler's optimization process. Registers are limited to 65536 per block on current hardware. This can be a limiting factor to the achievable level of parallelism: Kernels that require a lot of local variables can exceed this limit which means that the compiler either has to reduce the number of concurrent threads or place some of the local variables into slower memory regions. Regardless of grid and block configurations, current CUDA-capable processors always execute threads in clusters of 32 which are called *warps*. In theory, this means that the optimum occupancy rate can only be reached if the block size is a multiple of 32; however, due to the memory contraints, the highest possible performance does not necessarily conform to 100 per cent occupancy. Experimenting with block and grid sizes is a typical process of optimizing a CUDA application.

2.2.4. Alternative Parallel Programming APIs

The major drawback of CUDA is its limitation to the hardware of one specific vendor, NVIDIA. Especially in scientific applications open standards should generally be preferred over proprietary solutions. A GPU computing API that meets these requirements is the *Open Compute Library (OpenCL)*. It was specified by the Khronos Group, a consortium of hardware manufacturers also responsible for the de-facto standard library for hardware-accelerated graphics applications, OpenGL. The compute library is described in [Khronos OpenCL Working Group, 2015]: The concept of *hosts* and *devices*, execution model and memory model are very similar to that of CUDA. Threads, blocks and grids exist and work in the same way but are referred to as *work-items*, *work-groups* and *NDRanges*, respectively. The different memory regions of CUDA exist in OpenCL as well. Contrary to CUDA, OpenCL works on a wide range of different hardware platforms including GPUs of competing manufacturers as well as multi-core desktop and mobile CPUs. Despite this advantage CUDA was chosen for the purpose of the applications described herein; the reasons for this were better stability, availability of advanced debugging tools, higher performance and generally higher technological readiness at the time of commencement of this work. However, the software was designed in a way so that migration to a different API at a future time would be facilitated.

Another notable API for parallel programming is *OpenMP (Open Multi-Processing)*. Similar to OpenCL it is specified by a consortium of hardware and software manufacturers, the *OpenMP Architecture Review Board* ([OpenMP Architecture Review Board, 2013]). It works by extending C, C++ and Fortran compilers with special compiler directives that can be used in the source code in order to parallelize portions of the algorithm. For example, prepending a *for* loop with *#pragma omp parallel* in C causes each execution of the loop to be run in parallel on the available CPU cores. Other statements exist to specify which variables are shared and which are exclusive to each thread. *Barrier* directives are used to control thread synchronization. OpenMP was designed for use with multicore CPUs and high performance clusters; support for dedicated hardware such as GPUs was added only recently. OpenMP is very easy to use because parallelization can be achieved with only minor additions to the code; most C, C++ and Fortran compilers have built-in support for it so additional tools are not required. However, it does not possess the flexibility and optimization potential that dedicated GPU computing APIs offer. Another standard called *OpenACC (Open Accelerators)* aims to fill this gap by combining the ease of use of OpenMP with the versatility of OpenCL and CUDA ([OpenACC Group, 2013]).

2.2.5. GPU Computing in Space Research

Due to its widespread use and availability in source code form, the SGP4 algorithm described in [Vallado et al., 2006] has been a popular choice for experimenting with parallelized propagators. [Ahn, 2012] and [Fraire et al., 2013] describe implementations of the propagator in CUDA and OpenCL, respectively. Prior to that the SGP4 algorithm was ported to OpenCL in [Kebschull, 2011] as part of a parallelized numerical propagator. All sources show significant performance improvements over the original algorithm. The CUDA version for which the source code is available at [Ahn, 2012] has been adapted into a DOCTOR plugin and found to be real-time capable for larger populations ([Möckel et al., 2012]). Since the position error of SGP4 increases with propagation time ([Vallado et al., 2006] states an average error of 1 to 3 km per day) the algorithm is not suitable for long-term analysis. [Fraire et al., 2013] also discovered an additional position error introduced by using single precision floating point variables rather than double precision used in the original - about 12 km after 20 years. While insignificant within the error rate of SGP4, it shows that possible loss of precision is a problem that must be taken into account. This is especially true for GPU implementations where double precision operations are still significantly slower. As an example for an alternative application, [Hobson and Clarkson, 2012] present a method of using GPU computing for observation scheduling in the field of Space Situational Awareness (SSA). These publications hint to the fact that several people experiment with GPU computing in space research. However, apart from the SGP4 ports, no fully parallelized orbital propagator or comprehensive design guidelines for creating such an application have yet been published.

2.3. Software Architecture and Development

The following sections describe basic techniques of software development that are used in the context of this work. Concepts are presented using a C++ syntax where applicable but work similarly for other object-oriented programming languages.

2.3.1. Object-Oriented Programming Techniques

In software development, object-oriented programming has become the leading paradigm for medium-sized to large applications. The concept is an enhancement of the *procedural programming* paradigm which allows grouping sections of code into procedures, also called *subroutines* or *functions*. Each function is defined by a unique designator, a list of input parameters (which can be of length zero) and an optional return value. These items, collectively called the function's *signature*, can be used by other functions to execute and process the included code section. Fortran is a popular example of a procedural programming language.

Object-oriented programming languages expand on this concept by introducing structures called *classes* that are meant to represent real-world or abstract *objects*. Classes are sets of variables and functions (also called *attributes* and *methods* in this context) which form a logical unit. For example, a class representing a satellite could have the attributes *orbit* and *mass* which describe the object's properties, and a method called *propagate* which can be used to determine its current position. Classes can be *derived* from other classes which allows programmers to create more specialized versions of an object. Methods and attributes are then *inherited* from the parent class, i.e. they are available in the child class and do not have to be implemented again. It is however possible to reimplement an inherited method in a derived class; this process is called *overloading*. Child classes can be enhanced by adding further attributes and methods. In the above example a satellite could be natural, such as the Moon, or artificial. The *Satellite* class would be able to represent both so the classes *Moon* and *ArtificialSatellite* could both be derived from it. For the Moon, additional constants such as the escape velocity can be added as attributes while the artificial satellite might require an additional method to simulate the transmission of data which it collects. Since the lunar orbit is subjected to different physical influences than an artificial satellite, the *propagate* methods will likely have to be overloaded to account for those. In the design phase of object-oriented software, classes and their affiliation are commonly described in the *Unified Modelling Language* - a UML diagram for the above example is shown in figure 2.20.

Methods and attributes (as well as whole classes) can be classified using a variety of designators that define their *scope*, i.e. the range in which they are accessible to other functions. To prevent programming errors and unexpected behaviour the scope should be kept small, i.e. items should be accessible by as little other items as possible. The following shows a list of common designators and their effects.

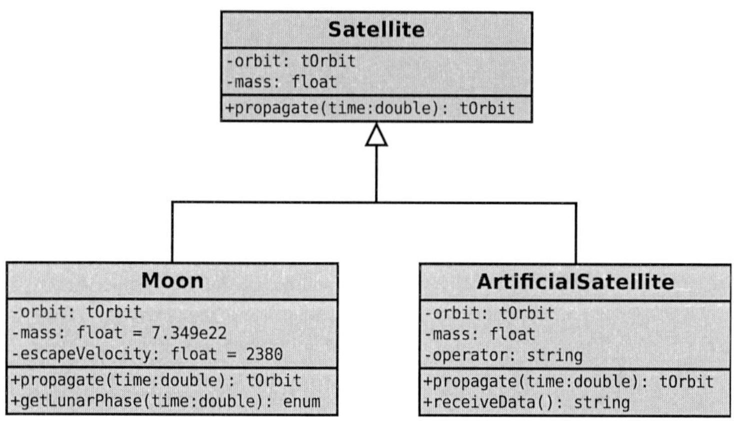

Figure 2.20.: UML diagram of the Satellite class and the two child classes derived from it.

- **public** items can be accessed by other functions outside of the object. All classes usually have a number of public functions that define their interface to the outside. In the UML diagram, these are marked with a "+" sign in front of their signature.
- **private** items can only be accessed by other functions inside of an object. All methods that are not meant specifically to be accessed by "outside" code should be made private. Usually, class attributes are also made private and public "setter" and "getter" functions are created to provide access to them. The reason for this is that such functions can check the given values for consistency before reading or writing to an attribute and can return an appropriate error message in case an invalid value was detected. In UML, private items are marked with a prefixed "-" sign.
- **protected** items are in an intermediate state between public and private ones. They can be accessed by the class that defines them as well as classes that inherit from it.
- **static** items are kept in a consistent state across all instances of an object; no matter how many instances exist of an object, if a static attribute is written in one of them, all others share this value. Static attributes can be used to store global information for which only one valid value exists such as the current system time.
- **abstract** or **virtual** functions are used to define empty methods that are meant to be overloaded by any class that inherits from the one containing the virtual. They are used to create classes that define interfaces but do not provide an actual implementation.

Grouping several classes into logical units can be done using *namespaces*, freely definable designators that are prepended to class names using a double-colon separator. For example, if a class called *Image* is defined within the namespace *Video* other classes which are outside that namespace must use the full name *Video::Image* to access it. Libraries often use namespaces to prevent naming conflicts with the applications that use them. In a similar manner, the software descriptions in the following chapters use a namespace notation to clarify which application the respective classes belong to.

2.3.2. Design Patterns

In software engineering, design patterns can be used during the architectural design phase. According to [Buschmann et al., 1996], the idea stems from the fact that certain problems occur frequently in many different software applications; a simple example would be "reading data from a file". For such problems, templates can be defined that describe the problem and provide an abstract solution; these templates are called *design patterns*. A popular example is the *Model-View-Presenter* pattern that is often used during the design of graphical user interfaces (GUIs). It separates an application's components into models, i.e. the data that is to be presented, views, i.e. the visual design of the GUI, and presenters that update the view according to changes in the model. Patterns are meant to provide an idea of how a specific problem can be solved; they do not provide functional software architectures. While simple patterns can sometimes be translated directly into an object-oriented class structure, doing so for complex ones often results in overly complicated architecture designs that are not tailored to the specific requirements of the software.

3 A Software Framework for Orbital Propagators

3.1. Properties of Orbital Propagators

After discussing the use cases (section 1.2.2) and mathematical properties (section 2.1) of analytical propagators, this section outlines the most apparent properties that such an application has from a software developer's point of view.

3.1.1. Complexity

Orbital propagators can have a wide range of complexity. In the most simple of all forms, a single line of code that calculates an object's mean motion from its orbital parameters can be regarded as a propagator. On the other side of the spectrum, a propagator can consist of a large variety of physical models simulating in great detail the different forces acting on the object. In any case, they are rarely meant to stand by themselves as single applications; instead, they are usually an integral part of other software tools designed to solve specific problems such as those detailed in section 1.2.2. The complexity of such superordinate applications can itself range from one-liners to huge software suites. With different knowledge and expertise required for the two components orbital propagators can and should always be developed independently from the application that makes use of them.

3.1.2. Modularity

Not only can the propagator be separated from its host application: As mentioned in the example above, an analytical propagator consists of several physical models describing various perturbation forces such as those listed in section 2.1.1. These models are developed by individual teams of researchers with different areas of expertise, collected and put together by other researchers who rely on orbital propagation for their work. [Vallado, 2007] states that modelling the atmospheric drag on a satellite alone requires expertise in the fields of molecular chemistry, thermodynamics, aerodynamics, meteorology and others. In many cases an algorithm is published in book or paper form and implemented individually by the people who use it. Over the years the models often get revised or corrected forcing all users to update their implementations. Sometimes an implementation in source code form is provided; however, directly integrating these into other applications is often impossible because it has been written in a different programming language or uses a different data format. In the commercial sector companies and individuals are usually hesitant to release applications in source code form because of copyright issues or fear of disclosing trade secrets.

3.1.3. Eligibility for Parallelization

Orbital propagation is often one of the most time-consuming components of an application. In the visualization use-case where propagated objects are merely displayed on a screen, the animation frame rate almost exclusively depends on how fast the objects' positions can be calculated. In long-term simulation, propagation is one of two critical components, the other one being conjunction analysis; therefore, optimization of only one of them has limited effects on the overall run time. As will be detailed in the following chapters, orbital propagation can be sped up to great lengths using GPU computing. This requires programming tools such as different compilers, libraries and knowledge of special programming languages and interfaces. Since the application using the propagator will likely be implemented without such techniques it should not have to rely on them in any way; the user of a propagator should only have to maintain the hard- and software required to run it, but not any tools or libraries that are necessary to build it.

3.2. Orbital Propagation Interface

3.2.1. Overview

The *Orbital Propagation Interface (OPI)* described in this chapter is the result of several years of research in the field of designing orbital propagators and involves several publications and students' theses. The basic idea first arose during the development of DOCTOR, the visualization software outlined in chapter 6. To create a fluent animation of space debris objects, a propagation algorithm was required that could calculate the positions of millions of objects within milliseconds. The first solution to this problem was to move the algorithm to the graphics processor in the form of a vertex shader. This is explained in more detail in [Möckel et al., 2011]. The major drawback of this method was that the resulting data was only available in the GPU's memory and could not easily be transferred back to the CPU; therefore, the improved algorithm could only be used for visualization but not for long-term simulations. This problem was solved by using CUDA instead of shader programs, with even better performance as shown in figure 3.1. These results demonstrated the opportunity of obtaining the same speedup for long-term simulation tools. Also in [Möckel et al., 2011], the first draft of an application framework, then called the *SPACE (Scientific Parallel Animation and Computation Environment) Framework*, was introduced. In their theses, [Rodermund, 2010] and [Lorefice, 2010] created the first proof-of-concept in the form of a visualization software and a simple CUDA propagator plugin, respectively, communicating over a common interface as shown in listing 3.1.

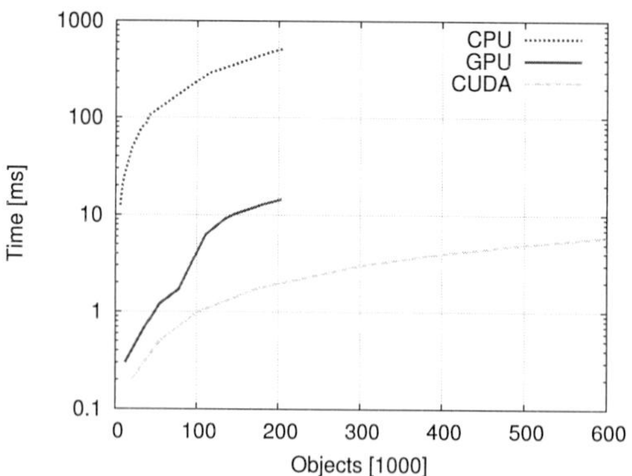

Figure 3.1.: Performance of the shader and CUDA code versus the CPU ([Möckel et al., 2011]). Results are combined from two experiments with different population sizes which is why the CUDA curve has more data points than the others.

Listing 3.1: First version of a propagator plugin interface ([Lorefice, 2010]).

```
#include "SimulationPlugin.h"
int main()
{
  //create simulator
  Propagator propagator;
  //initialize struct for orbit data on host
  SimulationData host(dtHost);
  //load 1000 orbit data sets from *.sim file
  Loader loader(std::string("data.sim"), host, 1000);
  //upload data
  simulator.upload(host);
  while(true)
  {
    double timeStep=10;
    //propagate
    propagator.run(timeStep, true);
    //download orbit data
    propagator.download(host);
    //visualization
    visualize(host);
  }
  //release data
  freeMem(host);
}
```

The interface was further refined in [Reglitz, 2012] based on a review of the three orbital propagators and their common parameters: FLORA ([Flegel, 2007]) which served as the basis for the parallel propagator described in chapter 4, FOCUS1 which is used in the MASTER Model ([Klinkrad, 2006]) and ZUNIEM, a numerical propagator that was originally created as part of a student's thesis ([Zuschlag, 1985]) and continuously enhanced over time. The work also provides an outline of a suitable software architecture taking into account a number of requirements such as modularity, support for parallelization and a flexible input/output component. Based on this research, the propagation interface and data structures were expanded further and published in [Möckel et al., 2012].

With the exception of combining area and mass into a single variable, the chosen values for orbital elements and object properties represent a minimal set from the recommended standard for orbital data messages proposed by [CCSDS, 2009] and can easily be extracted from data sets using that format. Finally, the original implementation of the Orbital Propagation Interface was carried out by [Thomsen, 2013]; some of the implementation-specific details have already been described in his work[1]. Since then, the software has been under constant development. The source code has been released under the GNU Lesser General Public License and is available online [Thomsen and Möckel, 2013].

3.2.2. Concept

To summarize the architectural requirements derived from the properties shown in section 3.1, an orbital propagator should be implemented in a way that

- keeps it separated from the application that uses it,
- allows different research groups to contribute individual parts, ideally in different programming languages,

[1]The software was originally called *Object Propagation Interface* but has since been renamed to clarify its purpose.

Parameter	FLORA	FOCUS1	ZUNIEM
Name of Computation	name	-	TEXT
Start Epoch	jd	IT0(1:5)	IT1...5,TIS
Semimajor Axis	el(1)	EL(1)	A
Eccentricity	el(2)	EL(2)	EPS
Inclination	el(3)	EL(3)	RIN
RAAN	el(4)	EL(4)	GOM
Argument of Perigee	el(5)	EL(5)	OM
Mean Anomaly	el(6)	EL(6)	PHI
Area to Mass Ratio	Am	PAR(2)	RKGPM2
Atmospheric Drag Coefficient	CD	PAR(1)	CD
Reflectivity Coefficient	CR	PAR(3)	RK
Propagation Time	time	IT(1:5)	ITAG, ISTUND, IMIN, SEK
Step Size	dt	DT	H

Table 3.1.: Excerpt from [Reglitz, 2012], table 3.18, showing the names of the input parameters that the three orbital propagators FLORA, FOCUS1 and ZUNIEM have in common.

- allows shipping and updating one part without affecting other parts of the system,
- allows one component to make use of advanced techniques such as GPU computing without causing dependencies for others,
- facilitates interchanging whole propagators, as well as individual physical models, at run time.

These requirements make orbital propagators an ideal candidate for a plugin-based implementation. [Marquardt, 1999] describes software patterns for plugins and plugin-based systems that are well suited for the problem at hand. According to the document, a *plugin* is defined as a piece of software designed to provide a specific functionality over a well-defined interface. An application that uses plugins to extend its functionality is called a *host*[2]. The plugin cannot be executed without a host but it can be developed, compiled and shipped independently. Typically, a plugin is implemented in the form of a dynamic library that is loaded by the host at run time. Depending on their function, host programs may rely on a plugin to be available or treat them as optional parts; for example, a plugin implementing a specific filter in a photo editing software is not crucial to the function of the application as a whole and therefore optional. On the other hand, most OPI hosts will likely rely on the availability of at least one propagator. Different plugin *types* can be supported; plugins are said to be of the same type when they share the same interface.

The basic concept of a software architecture using the plugin pattern is shown in figure 3.2. A host application capable of using plugins provides two components in addition to its core implementation. The first is the definition of the plugin's interface. In object-oriented programming, it is usually provided as a header file defining an abstract class with functions that the plugin needs to implement. A software component is only recognized as a plugin if all

[2] This is not to be confused with CUDA terminology in which the term *host* refers to the CPU-based platform, as opposed to the *device*, i.e. the GPU-based extension card.

necessary functions are provided so the host can expect them to be available. The second is the framework interface; it defines a set of functions in the host application that the plugin has access to, usually to gather required information or trigger specific actions. Together these two components define the ways in which the host and the plugin can communicate with one another.

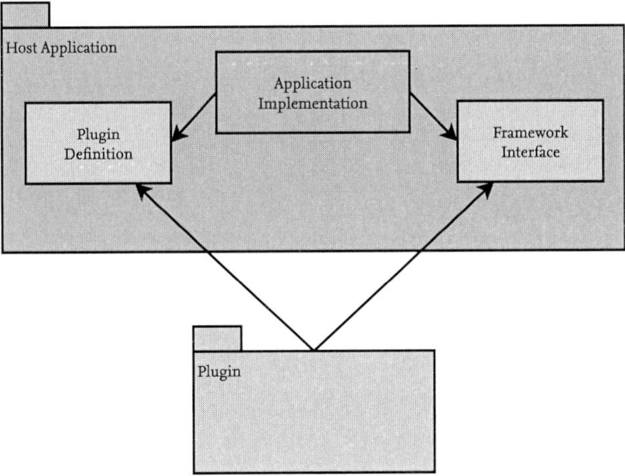

Figure 3.2.: Plugin Context as shown in [Marquardt, 1999].

OPI builds upon this basic concept but it does not suffice to achieve the goals outlined above: Not only are there many different propagators but also many potential host applications. Implementing a plugin interface into a single application will enable it to use any number of propagators that follow that specification. However, if another host was to gain access to this kind of flexibility, it would have to implement the interface on its own leading to undesired code duplication and possible inconsistencies. To resolve this issue, OPI was designed as a dynamic library that can be utilized by both host and plugin. It defines specific interfaces for both, thus acting as a framework between the two components. The basic structure is outlined in figure 3.3: Plugin definition and framework interface have been outsourced into the library; any host can use it to get access to the plugin interface.

The ability to provide compatibility with a multitude of host applications created for very different purposes poses the problem of designing an interface that is flexible enough to satisfy all their requirements. In the case of a single host application plugins simply abide by the standards specified by that software. On the other hand, a generic interface must be simple enough to be easily implemented into any host software and still be powerful enough to provide additional benefit. OPI only provides basic data types that are necessary to run an analytical propagator. If host or plugin require more information internally, these types can be used as a starting point to build elaborate structures around them. An example for this is given in chapter 6: In the visualization which serves as an OPI host, the population objects are represented by a class called *DOCTOR::SpaceObject* which contains orbital parameters in OPI's native format as well as additional data such as the objects' colors, names and coun-

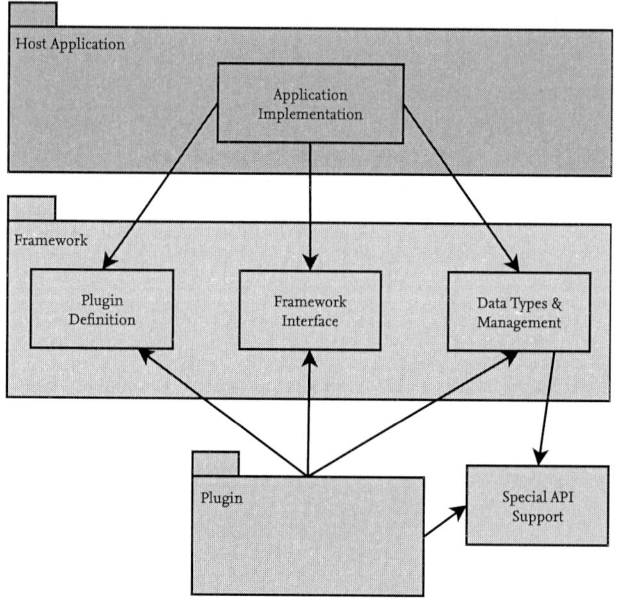

Figure 3.3.: The modified plugin design pattern upon which OPI is built.

tries of origin. Only the OPI-native data types are aggregated and sent to the propagator. OPI manages these data types automatically and provides a layer of abstraction that facilitates their use for the authors of both plugin and host. For further information that might have to be exchanged between host and plugin, OPI provides a mechanism called *PropagatorProperties* which is explained in section 3.2.6.

OPI was designed to support the use of GPU computing for propagators. As outlined in section 2.2.3, CUDA and other GPU computing APIs require complex memory management and manual copying of data between CPU and GPU memory. All functionality regarding CUDA is implemented into a separate component called "Special API Support" in the pattern that can be loaded at run time. This ensures that CUDA remains optional, both at compile time and at run time. If OPI is built with CUDA support, the library automatically checks for capable hardware on initialization. Plugins use a flag that indicates whether they rely on CUDA; if they do, they are only loaded if CUDA-capable hardware is found. The special API support component can be exchanged, for example, to provide the same automated mechanisms for OpenCL instead of CUDA.

Figure 3.4 shows the architecture derived from the pattern. OPI is designed as an object-oriented application and implemented in C++ (but can be used with other languages as described in section 3.2.8). OPI provides the classes *OPI::Host* and *OPI::Propagator* that serve as interfaces to the respective components. On the plugin side, OPI has additional classes such as *OPI::PropagatorModule* that can aid plugin authors in modularizing their code. Data management is handled via the *OPI::Population* class which can use the *OPI::CudaSupport* module to automatically handle memory transfers between the CPU and the GPU. The following sections describe OPI's components in detail.

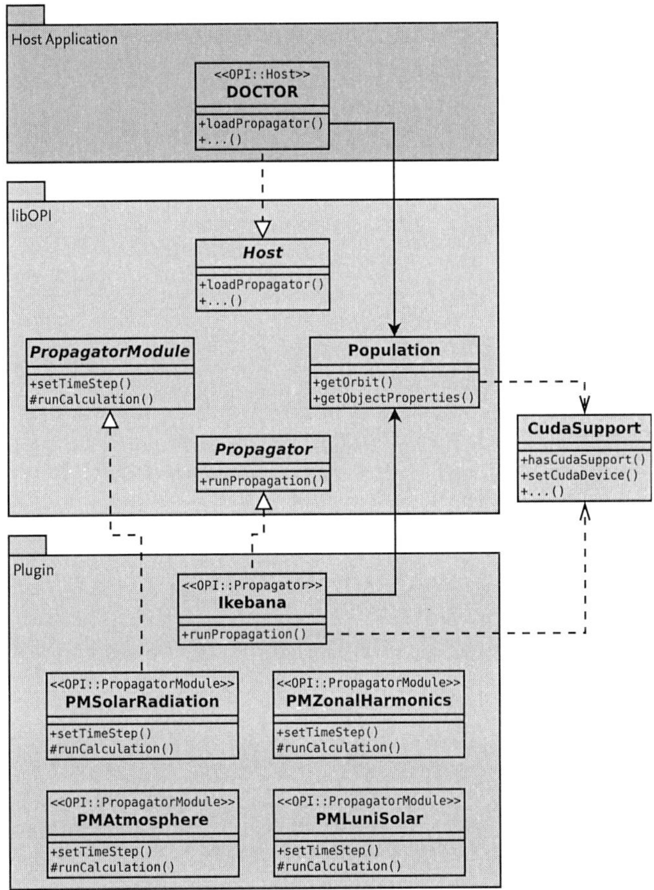

Figure 3.4.: UML diagram of the OPI interface connecting a host application (DOCTOR) and a propagator (Ikebana). For reasons of clarity, not all classes are shown.

3.2.3. Data Types

At the heart of OPI is the class *OPI::Population* which represents a population of objects orbiting around Earth. It serves as the main hub for data exchange between the host and the plugin. With parallel propagation in mind, this class is designed to manage an arbitrary number of objects. For the same reason, the data types representing the objects use single

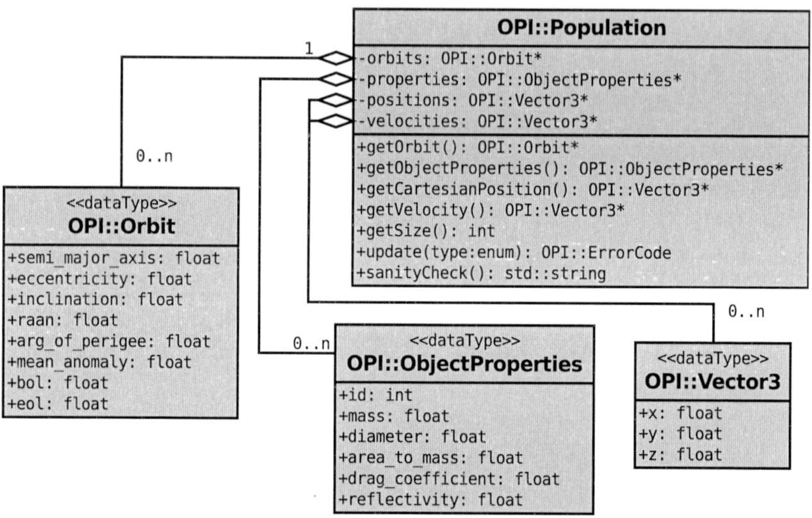

Figure 3.5.: Simplified UML diagram of *OPI::Population* and the associated data types *OPI::Orbit*, *OPI::ObjectProperties* and *OPI::Vector3*.

precision floating point variables only; as explained in section 5.4, current GPUs suffer from significant performance drops when double precision variables are used. The data types are *OPI::Orbit* which consists of the Keplerian elements as well as beginning of life (BOL) and end of life (EOL) dates, and *OPI::ObjectProperties* which includes additional information such as masses and reflectivity and drag coefficients (see table 3.2 for a detailed description). Another central data type is *OPI::Vector3* which is simply a three-dimensional vector that can be used for various information such as position and velocity in Cartesian coordinates. Figure 3.5 shows an UML diagram depicting the association between the *OPI::Population* class and the data types.

When an instance of *OPI::Population* is initialized, it starts with a population size of zero. Alternatively, the initial size can be specified as a parameter to its constructor, or set at a later point using the *resize* method. The constructor also requires a pointer to the host object as an additional argument. The class then generates vectors of the given size for orbits, object properties and Cartesian coordinates. One object of the population is represented by the elements of each of those vectors that share the same index. Cartesian coordinates must be generated by the propagator and are only available if the plugin provides support for them.

To fill the population instance with data, the methods *getOrbit* and *getObjectProperties* are used to provide pointers to the respective vectors (see listing 3.2 below). In case CUDA is used, the target device can be specified with an optional argument to these methods (either OPI_HOST for the CPU or OPI_DEVICE for the GPU). After changing the data, the *update* method needs to be called to notify OPI that new data is available. The *OPI::Population* class automatically

ObjectProperty	Description
id	A numerical ID for the object, for example, the NORAD ID for catalogue objects.
mass	The object's mass in kg.
diameter	The object's mean diameter in m.
area_to_mass	The object's front-facing area in square metres divided by its mass in kg. Sometimes the inverse, the mass-to-area ratio is used.
drag_coefficient	The object's drag coefficient. A default value of 2.2 is often used. OPI leaves it uninitialized so it must be set by the host.
reflectivity	The object's reflectivity coefficient. A default value of 1.3 is often used. OPI leaves it uninitialized so it must be set by the host.

Table 3.2.: Description of the elements in *OPI::ObjectProperties*.

handles the data transfer between the CPU and the GPU via the *OPI::CudaSupport* class which is explained in section 3.2.7.

3.2.4. Host Interface

OPI::Host
+loadPlugins(directory:std::string): OPI::ErrorCode +getPropagator(index:int): OPI::Propagator* +getPropagator(name:std::string): OPI::Propagator* +hasCUDASupport(): bool +selectCudaDevice(deviceNumber:int): int

Figure 3.6.: Simplified UML depiction of *OPI::Host*, the class that specifies the host interface.

The class *OPI::Host* represents the OPI's host-side interface. It provides methods allowing the host to initialize and choose from the available plugins, set the population data and run the selected propagator. As mentioned before, all device memory transfers for CUDA are handled automatically by OPI. This means that the author of the host requires no knowledge of parallel programming. If CUDA capable hardware and drivers are available, CUDA-enabled plugins will still function. The host has the options to check a chosen plugin for CUDA support and select a device if multiple GPUs are present using the appropriate methods.

To use an OPI propagator, the host application initializes an instance of the *OPI::Host* class or derives one of its own classes from it. The next step is to call the method *loadPlugins* which takes as argument the name of a directory in which the plugins can be found. This directory is traversed and identified propagators are loaded. Plugins requiring CUDA are skipped automatically if no capable hardware was found. To initialize population data, the host simply creates an instance of the *OPI::Population* class and associates it with the host object (see listing 3.2). Like explained in the previous section, the methods *getOrbit*, *getObjectProperties* and *getSize* can be used by the host to iterate through the uninitialized objects and assign values to each orbital element and object property.

Listing 3.2: Excerpt from a C++ program initializing an OPI host and some population data.

```
// Initialize a host object
OPI::Host host;

// Load plugins from the given directory
host.loadPlugins("plugins");

// Create a population of 200 objects and associate with the host object.
// This must be done after loading the plugin directory!
OPI::Population data(host, 200);

OPI::Orbit* orbits = data.getOrbit(OPI::DEVICE_HOST);
OPI::ObjectProperties* props = data.getProperties(OPI::DEVICE_HOST);

// Set data for orbital elements and properties
for (int i=0; i<data.getSize(); i++) {
  orbits[i].semi_major_axis = 7800.0f;
  //... continue with other elements

  // Do not forget these two!
  props[i].reflectivity = 1.3f;
  props[i].drag_coefficient = 2.2f;
  //... continue with other properties
}

// Notify OPI of the population changes
data.update(OPI::DATA_ORBIT);
data.update(OPI::DATA_PROPERTIES);
```

3.2.5. Plugin Interface

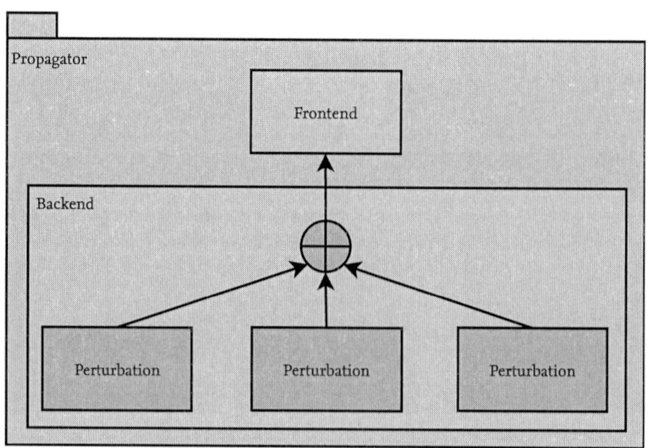

Figure 3.7.: Design pattern for an analytical propagator.

The architecture for OPI's plugin side is based on a simple design pattern that is derived from the natural properties of orbital propagation physics. The pattern is shown in figure 3.7. It

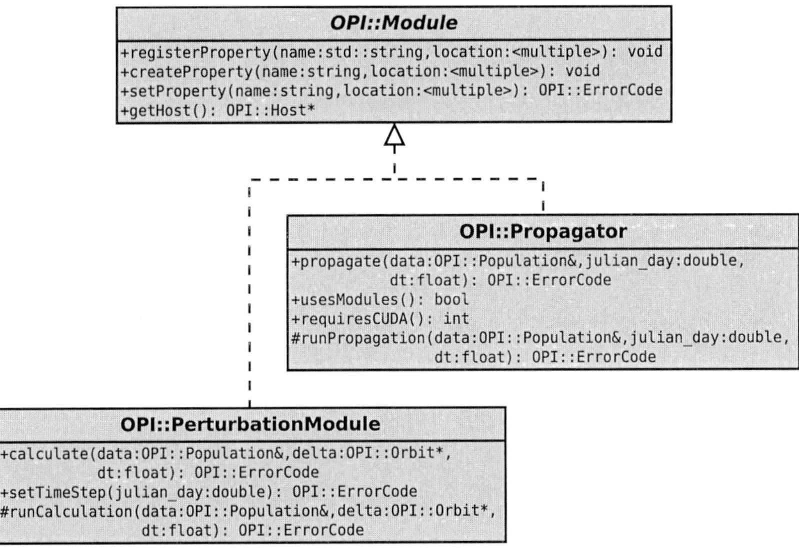

Figure 3.8.: Simplified UML diagram of *OPI::Propagator* and *OPI::PerturbationModule*, both derived from the *OPI::Module* class.

consists of two main components, a frontend and a backend. The frontend communicates with the application that uses the propagator, provides input data, returns results and manages the execution of the plugin components. The backend contains the actual propagation algorithms. It is split into a number of "perturbation" components that share the same interface. Each of these components individually calculates a perturbed orbit based on the original orbit, current time and step size, as well as additional information such as the object's mass, reflectivity and drag coefficients. The perturbed orbits are aggregated and handed to the frontend.

In this simple case it is possible to translate the pattern directly into an object-oriented software architecture. OPI provides two classes for this shown in figure 3.8: *OPI::Propagator* for the frontend and *OPI::PerturbationModule* for the backend. Both are derived from a superclass *OPI::Module* which provides features common to both parts such as the *PropagatorProperties* described in section 3.2.6. *OPI::Propagator* is the main class that a propagator needs to implement to be recognized as an OPI plugin. Its public methods are exposed to the host program and called to perform the propagation. The most important one is the *propagate* function which takes as arguments an instance of *OPI::Population* to which the host has previously written the objects' orbits and properties as well as the propagation time and step size. The propagation results are written back to the *OPI::Population*. Usually, this function is called by the host in a loop over several time steps so the next step continues with the updated population from the last one (see listing 3.3). The *propagate* function is a wrapper function that checks whether data needs to be transferred between CUDA devices prior to calling the actual propagation function, *runPropagation*. This is the one that plugin authors override to implement the actual functionality, as shown in listing 3.4. Another public function that should be implemented is *requiresCUDA* which returns an integer corresponding to the minimum CUDA version that this plugin requires in order to run. If not overridden, the default value is zero signalling no dependence on CUDA. Also shown in the listing are several preprocessor macros supported by OPI that can be used to set general plugin information such as name, author, description and version number.

Listing 3.3: Excerpt from an OPI host running a propagation plugin.

```
// Initialize host and population data as shown in the previous example:
OPI::Host host;
host.loadPlugins("plugins");
OPI::Population data(host, 200);
//...

// Fetch pointer to a specific propagator
OPI::Propagator* ikebana = host.getPropagator("Ikebana");

// Set initial date (Julian day) and step size (one day = 86400 seconds)
const double julianDay = 2440980.0;
const float stepSize = 86400.0f;

// Check if the propagator exists
if (ikebana) {
  // Propagate for one year, starting at julianDay, with the given step size
  for (int t=0; t<365; t++) {
    double currentTime = julianDay + t;
    ikebana->propagate(data, currentTime, stepSize)
  }
}

// Fetch pointer to updated orbital elements
OPI::Orbit* results = data.getOrbit(OPI::DEVICE_HOST);
```

Listing 3.4: A simple OPI propagator in C++.

```
#include "OPI/opi_cpp.h"
#include <iostream>

// Set propagator name and version
#define OPI_PLUGIN_NAME "Example Propagator"
#define OPI_PLUGIN_AUTHOR "M. Moeckel"
#define OPI_PLUGIN_DESC "A simple example"
#define OPI_PLUGIN_VERSION_MAJOR 0
#define OPI_PLUGIN_VERSION_MINOR 1
#define OPI_PLUGIN_VERSION_PATCH 0

// Derive from OPI::Propagator
class ExamplePropagator: public OPI::Propagator
{
  public:
    // Implement functionality into runPropagation method
    virtual OPI::ErrorCode runPropagation(
      OPI::Population& data, double julian_day, float dt )
    {
      // Fetch pointer to orbital data
      OPI::Orbit* orbit = data.getOrbit();

      // Add mean motion to each orbit's mean anomaly
      for (int i=0; i<data.getSize(); i++) {
        const float meanMotion = sqrt(MUE / pow(orbit[i].semi_major_axis,3.0f));
        orbit[i].mean_anomaly += meanMotion;
      }

      // Notify OPI of the updated orbits
      data.update(OPI::DATA_ORBIT);

      // Return successfully
      return OPI::SUCCESS;
    }

    // Notify OPI that this propagator does not require CUDA
    virtual int requiresCUDA() { return 0; }
};

// Include additional functions that define the plugin
#define OPI_IMPLEMENT_CPP_PROPAGATOR ExamplesPropagator
#include "OPI/opi_implement_plugin.h"
```

While it is possible for very simple propagators to implement all functionality into a single class derived from *OPI::Propagator* it is recommended to divide the plugin into a frontend and a backend with the latter containing individual classes for the perturbation models. Each of them can be derived from the *OPI::PerturbationModule* class that provides a suitable interface. This means that perturbation modules can be developed independently by different authors; the common interface makes it easy for others to collect different modules and combine them into a working propagator. Implementations of *OPI::PerturbationModule* can also be compiled into stand-alone sub-plugins. By using the *OPI::CustomPropagator* class instead of *OPI::Propagator* a propagator can be constructed at run time from a choice of individual, pre-compiled perturbation modules. Thus, updating a propagator or an individual model can simply be achieved by replacing a shared object in the plugin folder.

The interface of *OPI::PerturbationModule* works similarly to that of the frontend class. The most important method to be implemented is *runCalculation*. It takes as argument the current

Population, a pointer to an array of *OPI::Orbits* with the same size as the population, and the current time step. The function serves as the entry point for running a perturbation module's private methods that implement the respective physical model. The current time is set in a different function, *setTimeStep* which takes as input the respective Julian date. The reason for having two individual functions will become apparent in chapter 4, where a CUDA propagator using OPI is introduced: The *setTimeStep* function allows the plugin to precalculate certain values that depend on the current time but not on the objects' properties (e.g. Sun and Moon positions in third body perturbation). If such values are computed when the time step is set and stored in class variables for later access, a lot of computation time can be saved compared to running through those same calculations for every object. Like in *OPI::Propagator*, there is a public wrapper function called *calculate* with the same signature as *runCalculation* that checks for and performs necessary data transfers before calling the actual calculation. An example of an OPI propagator using a PerturbationModule is shown in listing 3.5.

Listing 3.5: The *runPropagation* function of a simple OPI propagator using a PerturbationModule.

```
// It is assumed that an instance of a perturbation module has been created inside the
// class and that memory for resulting orbits has been allocated.
OPI::PerturbationModule perturbation;
OPI::Orbit* delta;

// Implementation of runPropagation using a perturbation module.
OPI::ErrorCode runPropagation(OPI::Population& data, double julian_day, float dt )
{
  // Fetch pointer to orbital data
  OPI::Orbit* orbit = data.getOrbit();
  // Initialize orbit array for perturbation
  OPI::Orbit delta[data.getSize()];
  // Forward current time to perturbation module
  perturbation.setTimeStep(julian_day);
  // Run perturbation module, results will be stored in delta
  perturbation.calculate(data, delta, dt);
  // Add deltas to original orbits
  for (int i=0; i<data.getSize(); i++) {
    orbit[i] += delta[i];
  }
  // Notify OPI of the updated orbits
  data.update(OPI::DATA_ORBIT);
  // Return successfully
  return OPI::SUCCESS;
}
```

3.2.6. PropagatorProperties

One of the drawbacks of having a generic interface is that it can never account for every kind of information that might need to be exchanged between a host and a plugin. The OPI interface has been designed to provide the lowest common denominator of the examined propagators. However, the physical models for some of the perturbation forces will most likely rely on additional data. In some cases, such data can be included with the plugin and loaded on initialization, for example, a file containing an atmospheric density table. In other cases, it might be provided by the host (or the user via the host application's graphical user interface). This can be achieved in OPI by using a mechanism called *PropagatorProperties*. A PropagatorProperty is a variable of any supported type that is defined and allocated inside the plugin. By using the *registerProperty* function as shown in listing 3.6, the plugin author can make a pointer to this variable available to the host under a specified designator. The host application can then set a value to this variable by supplying it, along with the designator, to the *setProperty* function[3].

Supported variable types for PropagatorProperties are *float, double, integer* and *string* as well as arrays of the three numerical types. PropagatorProperties are implemented in the *OPI::Module* superclass which means that not only propagator plugins can have properties, but individual *OPI::PerturbationModules* can use them as well. As an alternative to *registerProperty*, the method *createProperty* can be used in the same manner, with the difference that OPI will manage the associated variable internally so the module does not need to explicitly allocate it. In addition to sending data, PropagatorProperties can also be used to set configuration options such as activation or deactivation of a specific perturbation force at run time.

Listing 3.6: Excerpt from an OPI plugin that registers PropagatorProperties upon initialization and sets default values.

```
// allocate variables for properties and set sensible default values
int opt_useAtmosphere = 1;
int opt_useThirdBody = 1;
int opt_useGravity = 1;
int opt_useSolarRadiation = 1;

// register OPI properties that can be configured by the host
registerProperty("useAtmosphere", &opt_useAtmosphere);
registerProperty("useThirdBody", &opt_useThirdBody);
registerProperty("useGravity", &opt_useGravity);
registerProperty("useSolarRadiation", &opt_useSolarRadiation);
```

The obvious drawback of this method is that the plugin may rely on data that is defined outside the interface. Although the host can retrieve a list of PropagatorProperties from the plugin as shown in listing 3.7 it can never provide optimal support for all possible properties of all existing plugins. In the listing, for example, the host loads the propagator "Ikebana" and sets its property "useAtmosphere" to zero which implies that the author of the host application is aware of the existence of that property. This is bad practice as that code only works for one specific propagator, and it would fail if the plugin author were to remove or rename that property. A better solution would be to present the available properties to the user via a graphical interface or a configuration file. An example for such a file is outlined in listing 3.8: Its syntax provides a very easy way of setting the properties for every propagator in use and can be parsed by the host using the C++ code shown in listing 3.9.

[3][Thomsen, 2013] points out that for a property to be visible to the host, the *registerProperty* function must be called at initialization, i.e. in the class constructor (C++) or by implementing and calling it from the *OPI_Plugin_Init* function (C/Fortran).

Listing 3.7: Excerpt from a host application that prints all properties of a specific plugin and sets a known value.

```
// Initialize host and population data as shown in the previous example:
OPI::Host host;
host.loadPlugins("plugins");
OPI::Population data(host, 200);
//...

// Fetch pointer to propagator "Ikebana"
OPI::Propagator* ikebana = host.getPropagator("Ikebana");

if (ikebana) {
  // Print a list of the propagator's available properties
  for (int i=0; i<ikebana->getPropertyCount(); i++) {
    std::string name = ikebana->getPropertyName(i);
    std::cout << name << ": " << ikebana->getPropertyString(name) << std::endl;
  }

  // Set the property "useAtmosphere" to zero
  ikebana->setProperty("useAtmosphere", 0);
}
```

Since the PropagatorProperties do not show up in the interface definition it is extremely important that plugin authors keep the amount of properties to a minimum, and those that are there need to be well documented. In the plugin code, the properties should assume sensible default values that allow the plugin to run even if no options are set. Any information that is vital for the plugin to run should be aquired by different means, e.g. loaded from a file provided with the plugin. A PropagatorProperty can be used to provide an optional means to override this default. For the further development of OPI, existing plugins should be regularly analyzed and common properties should be added as core interface functions.

Listing 3.8: Example of a configuration file for user-definable PropagatorProperties.

```
# Here you can set properties for individual propagator plugins in the following form:
# PropagatorName:Property=Value
# Values can be strings, integers and floats. Strings are identified by enclosing them
# in quotation marks, i.e. Propagator:Property="Value"
# Floats are identified by the decimal dot, i.e. Propagator:Property=1.0
# Everything else is parsed as an integer.

# Ikebana options
# toggle zonal harmonics perturbations (1=on, 0=off)
Ikebana:useGravity=1
# toggle solar radiation pressure perturbations (1=on, 0=off)
Ikebana:useSolarRadiation=1
# toggle third body perturbations of the sun and moon (1=on, 0=off)
Ikebana:useThirdBody=1
# toggle atmospheric drag perturbations (1=on, 0=off)
Ikebana:useAtmosphere=1
# set zonal harmonics J-Terms (2, 4 or 6)
Ikebana:jTermLevel=6
# set verbose output (0=none, 1=default, 2=more, 3=lots, 4=extreme)
Ikebana:verboseLevel=1
# toggle generation of objects' Cartesian positions (1=on, 0=off)
Ikebana:cartesianPosition=0
```

Listing 3.9: OPI host code that is able to parse the above configuration file and set the PropagatorProperties accordingly (error handling and definition of some functions omitted for clarity).

```
// get next line from the config file:
string line = configFile.getNextLine();
// first, split it at the colons
vector<string> tokens = tokenize(line,":");
// proceed if a colon was found
if (tokens.size() == 2) {
  // use the part before the colon as the propagator's name
  string propagatorName = tokens[0];
  // split the remaining line at the equals sign
  vector<string> property = tokenize(tokens[1],"=");
  // proceed if the split was valid
  if (property.size() == 2) {
    // set the first part as property name...
    string propertyName = property[0];
    // ...and the second as the value
    string propertyValue = property[1];

    // get a pointer to the propagator with the given name
    OPI::Propagator* p = host->getPropagator(propagatorName);
    // proceed if a propagator with that name exists
    if (p != NULL) {
      // if the value is enclosed in quotes, parse it as a string
      if (propertyValue[0] == '\"' && propertyValue[propertyValue.length()-1] == '\"') {
        p->setProperty(propertyName, propertyValue.substr(1,propertyValue.length()-2));
      }
      // otherwise, if the value contains a dot, parse it as a float
      else if (propertyValue.find_first_of('.') != string::npos) {
        p->setProperty(propertyName, stof(propertyValue));
      }
      // parse everything else as an integer
      else {
        p->setProperty(propertyName, stoi(propertyValue));
      }
    }
  }
}
```

3.2.7. CUDA Support

OPI::CudaSupport
+getDeviceCount(): int +selectDevice(device:int): void +allocate(a:void**,size:size_t): void +copy(a:void*,b:void*,size:size_t,host_to_device:bool): void +free(mem:void*): void

Figure 3.9.: Simplified UML diagram of the *OPI::CudaSupport* class.

The Orbital Propagation Interface was designed to facilitate the use and creation of GPU-based propagators. Plugin authors should be able to parallelize their propagators with minimal effort while host authors should be able to use GPU propagators without any additional knowledge of the technology. Since the use of CUDA is optional and other GPU computing libraries exist that can be supported in the future, all CUDA-related code is placed inside a separate class called *OPI::CudaSupport.* This class can be deactivated at compile time; this way, OPI does not rely on the CUDA SDK to be built if GPU computing capabilities are not required. To support a different GPU computing library such as OpenCL, the class can simply be exchanged for an equivalent one that replaces all CUDA functions with their respective OpenCL counterparts.

The *OPI::CudaSupport* module is used by *OPI::Host* and *OPI::Population.* Figure 3.9 shows its most important functions. The host uses the *getDeviceCount* function to check for compatible devices at run time. If none are found (or the module does not exist at all) GPU-based propagators are skipped during the search for plugins. Otherwise, the *selectDevice* function can be used to choose from a list of devices if more than one compatible GPU is available. For the *OPI::Population,* the module provides the functions *allocate, copy* and *free.* They are used to reserve or free GPU memory and copy data between (CUDA) host and device. These functions simply act as wrappers for the respective CUDA calls (*cudaMalloc, cudaMemcpy* and *cudaFree*) and can easily be replaced to support other GPU computing libraries.

3.2.8. Multi-Language Support

One of the key drivers behind the design of OPI was the ability to combine modules written in different languages. This improves both ease of use because every developer can use their preferred language, and cooperation because it makes it simpler to adapt existing code. OPI is written in C++ and its design draws strongly from object-oriented principles. Host and plugin using the framework can adapt this by deriving from OPI's classes as shown in the above examples. However, it is also possible to use a different approach. Two of the most frequently used programming language in space research are Fortran and C. OPI provides interfaces to both of these languages. They are generated automatically from the C++ code during the build process and translate OPI's functions into appropriate equivalents in the target languages. Techniques specific to object-oriented programming are slightly modified; for example, instead of creating an instance of *OPI::Host* or deriving a class from it, C and Fortran host applications use the function *OPI_create_Host* which creates a host object that is maintained by OPI internally and returns a pointer to it. This pointer is then passed on to other function calls that require access to it. On the plugin side, the Fortran and C interfaces simply provide a list of functions that need to be implemented; this works just like the C++ version except that the functions are not grouped in a class. Since host and plugin are independent from each other they can be developed in different programming languages and still work together as long as they conform with the interface definition. Many other programming languages such as Java and Python as well as tools like Matlab provide support for accessing C libraries. This can be used to extend OPI's range of supported platforms even further. For example, the host interface can be accessed from Matlab to call a propagator plugin and retrieve its output.

3.2.9. Collision Detection

Besides orbital propagation, OPI supports two other plugin types for assessing collision risks of populations. It was developed as part of the research in [Thomsen, 2013]; it works independently from the propagation functionality and is only mentioned here for completeness. One of the plugin types is used for spatial partitioning, the other for calculating the actual collision probabilities. An example is shown in figure 3.10: Here, the space around Earth is partitioned into cubes which are progressively subdivided until only two objects remain inside one cube. The resulting list of possible collision pairs is used as input to the second plugin which calculates the cumulated collision risk based on a method published in [Liou, 2006]. The partitioning and risk assessment are executed once per time step, i.e. several times per second in the visualization. Both plugins have access to the CUDA support module so they can be implemented using GPU computing. Thomsen found that the benefits obtained through parallelization depend largely on the sorting algorithm as well as the nature of the population.

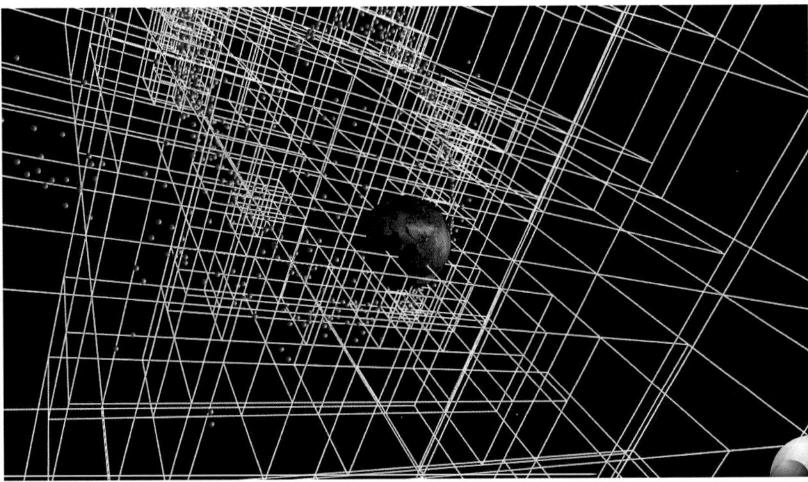

Figure 3.10.: Visualization of the real-time spatial partitioning and collision risk assessment in DOCTOR.

3.3. Propagator Implementation Guidelines

The plugin pattern describes the necessity for a common interface. This is essential in an environment that requires easy interchangeability and contributions from multiple authors. It is, however, not sufficient to ensure seamless integration of a propagator into a host application. Additional guidelines specific to orbital propagators must be defined to ensure that all propagator plugins work as expected. Some of these guidelines can be implemented into the plugin framework; but as strong enforcement of regulations might come at the cost of flexibility or execution speed that decision must be made with great care. This section gives the most important guidelines that host and plugin authors shall abide by to ensure compatibility. All guidelines include the following information:

- **Scope** indicates whether this rule applies to host or plugin authors (or both),
- **Rationale** explains the reason why this rule should be followed,
- **Exceptions** lists possible exceptions that need to be kept in mind,
- **Enforcement** explains whether, and how, this rule can be enforced in OPI.

If OPI is updated in the future, the rule set has to be revised to account for the changes if necessary. Missing regulations should be added whenever uncertainties and incompatibilities are discovered during practical use.

Rule 1	**Only valid orbital data shall be provided over the interface.**
Scope	Host and Plugin
Rationale	Propagator and host application are two separate software projects that are potentially developed by different people in different languages. In order for both components to work together each of the components has to fulfill certain expectations on which the other can rely. This includes taking responsibility for conforming to the standards set at interface level. Valid value ranges for orbits and object properties are listed in tables 3.3 and 3.4, respectively. Default values are given as a suggestion and are not set automatically. Semi major axes smaller than 6400 km can be ragarded as invalid and affected objects should be marked as decayed. Beginning of life and end of life dates before 1950 can also be disregarded since no man-made objects existed in space prior to that date.
Exceptions	Some propagators may allow hyperbolic orbits ($e >= 1$) or dates earlier than 1950 when natural objects are propagated. This should be explicitly stated in the documentation.
Enforcement	None. While the providing application should be responsible for correctness, the receiving application should apply sensible checks (e.g. whether drag and reflectivity coefficients are set).

Value	Valid Range
Orbit.semi_major_axis	≥ 6400
Orbit.eccentricity	$0 < \epsilon < 1$
Orbit.inclination	$0 \leq i \leq 2\pi$
Orbit.raan	$0 \leq \Omega \leq 2\pi$
Orbit.arg_of_perigee	$0 \leq \omega \leq 2\pi$
Orbit.mean_anomaly	$0 \leq M \leq 2\pi$
Orbit.bol	0 or $\geq$ MJD1950
Orbit.eol	0 or $\geq$ MJD1950

Table 3.3.: Constraints for orbital data provided by the host

Value	Valid Range
ObjectProperties.id	32-bit integer
ObjectProperties.mass	> 0
ObjectProperties.diameter	> 0
ObjectProperties.area_to_mass	> 0
ObjectProperties.reflectivity	$0 \leq C_R \leq 2$ (default: 1.3)
ObjectProperties.drag_coefficient	> 0 (default: 2.2)

Table 3.4.: Constraints for object properties provided by the host

Rule 2	**Angles shall always be given in radians.**
Scope	Host and Plugin
Rationale	Practical experience shows that a lot of programming errors arise from interpreting a degree angle as radians and vice versa. Therefore, one of them should be chosen as a standard to be used throughout the application. Since the intrinsic trigonometric functions of virtually all programming languages use radians, this is the most sensible choice as it prevents unneccessary conversions.
Exceptions	May arise when angle values are provided in degrees or the host application requires degrees for file input/output. In these cases, values should be converted directly after reading and before writing, respectively. Should it be necessary to write a degree value to a variable first the name of that variable should reflect that the value is given in degrees.
Enforcement	Not trivial since a number smaller than 2π can be both radians or degrees. Instead, a debug function is provided by OPI that allows to check all angles and generate a warning if a number larger than 2π is detected.

Rule 3	**Dates shall be given in the Julian Date (JD) format.**
Scope	Host and Plugin
Rationale	Many physical models rely on date and time information which are usually given in the Julian Date format. As an alternative, the Modified Julian Date is commonly used in astronautics applications because it provides the ease of use of the regular Julian Date but prevents unnecessarily large numbers by omitting the days before space flight was practiced. It is defined to commence at midnight, January 1st, 1950, hence it is offset from the regular Julian date by 2400000.5. However, some models might not be able to handle this restriction, i.e. if natural objects are regarded or information from earlier dates is required. The Modified Julian Date can be used internally by the plugin but should be converted before it is sent over the interface.
Exceptions	No exceptions.
Enforcement	Cannot be enforced without limiting the use of OPI to certain time periods.

Rule 4	**A plugin shall never remove objects from the population.**
Scope	Plugin
Rationale	An object that has decayed is of no use for the plugin and can therefore be excluded. However, since the author of a propagator does not know how the propagation results are used they must not remove information that the host might rely on. For example, the host might seek through the result set based on previously stored indices or try to access the last known position of a decayed object. In general, a plugin should always add or change information but never remove any. Decayed objects shall be marked accordingly - see next rule.
Exceptions	No exceptions.
Enforcement	Due to the structure of OPI this rule cannot be enforced directly. However, it is easy to keep in mind and follow in its slightly rephrased form: *The plugin shall never call OPI::Population's resize function.*

Rule 5	**Decayed objects shall be marked and left unchanged.**
Scope	Plugin
Rationale	Information about whether an object has decayed is crucial to many use cases of propagators. It is therefore important that hosts can rely on the plugins to supply that information. In OPI, the *OPI::Orbit.eol* tag must be set to the Julian date of the current propagation step to mark a decay. Objects that are marked as decayed should not be propagated further to preserve information about the decay orbit and to save computational time.
Exceptions	No exceptions.
Enforcement	None. However, a future version of OPI could enforce this rule by automatically marking decayed objects either in *OPI::Population* or *OPI::Module*.

Rule 6	**A plugin shall never return a NaN value.**
Scope	Plugin
Rationale	A host must be able to rely on the plugin to provide reasonable results. NaN (*not a number*) values are usually a result of improper error handling in the propagation algorithm and therefore do not fall into the host author's responsibility. When a NaN is generated it often propagates through the following calculations invalidating their results and making it difficult to spot their origin. Therefore, they should be caught directly at the place of occurrence.
Exceptions	If NaN values are produced as a result of the host providing invalid input data they do not fall into the plugin author's responsibility. Still, the possibility of such errors occurring should be kept in mind and dealt with appropriately.
Enforcement	Automatic NaN checks can be implemented into the framework at certain points, e.g. before returning an *OPI::Orbit* to the host. As this process might be very time-consuming, NaN checks are implemented only in a "sanity check" routine used for debugging. Plugin authors need to identify possible NaN occurrences in their algorithms and make sure they are properly handled. Depending on the nature of the plugin, generic NaN checks can be implemented at an appropriate place where no significant performance reductions are expected.

Rule 7	**The host should provide a means for the user to configure Properties.**
Scope	Host and Plugin
Rationale	Since a plugin can freely define any number of properties, it is impossible for a host application to take them all into account. The host should therefore be able to present to the user a list of properties for the chosen plugin combined with a method of letting the user choose sensible values. Property names shall not be hard-coded into the host application. In turn, plugins should assume reasonable default values for all of their properties.
Exceptions	This rule can be omitted if every functionality required by the host is covered by the standard interface.
Enforcement	None, since the preferred way of setting properties may vary between host applications (e.g. GUI mask or text files). The example code shown earlier in this section can be used as a template for implementing text file based property setup into host applications. However, a default method of setting properties can be implemented in a future version of OPI.

4 High-Performance Analytical Propagation

In this chapter two analytical propagators are introduced that are implemented as OPI plug-ins. The first is FLORA (Fast, Long-term Orbit Analysis), a Fortran 90 application that has been developed at the Institute of Space Systems. It has been continuously improved over the years for stability, speed and accuracy, the latest addition being the implementation of the OPI interface. The second is Ikebana, a port of FLORA written in C++ and CUDA. It has been created to prove the feasibility of GPU computing for the task of orbital propagation, with the aim to recreate the output of FLORA as accurately as possible while exploiting the GPU's parallelism for run time improvements. The work has been conducted in the context of [Möckel et al., 2015] where preliminary results of Ikebana's performance analysis have been published.

4.1. FLORA

4.1.1. Overview

FLORA is an analytical propagator developed at the Institute of Space Systems. As the name suggests it was designed with an emphasis on long-term analysis of orbital populations. It takes into account perturbations from atmospheric drag, zonal harmonics, solar radiation pressure as well as gravitational pull from Sun and Moon. Their implementations are based on the analytical models described in section 2.1.1. Short-periodic perturbations are generally omitted as they do not have any effect on the population's long-term development. Each perturbation model is implemented in its own subroutine with an individual interface. These subroutines are executed subsequently in each time step and return perturbed orbits which are added to the original orbit, along with the object's mean motion during that time, after all models have been run. Original orbital data and object properties are stored in global variables so they can be accessed by all subroutines. Double precision variables are used for all floating point values. With the exception of the atmospheric model, the perturbation forces are very straight-forward implementations of the analytical algorithms described in section 2.1.1. Further details of the implementation are given in [Flegel, 2007] with updates to the atmospherical model described in [Radtke, 2011] which are outlined in the following sections.

FLORA can be operated in two modes. As a standalone application, it reads orbital data, object properties and propagation time for a single object from a given input file and outputs the resulting orbit in each time step. Further configuration options in the same input file can be set to disregard certain perturbation models. Alternatively, FLORA can be integrated into other applications which can then use the interface function shown in listing 4.1 to execute a propagation step. FLORA is not designed as a dynamic library; integration is performed by simply linking the object codes of the two applications into a single executable. FLORA supports propagation of only one object at a time. To process multiple objects, the whole application has to be executed once for each object. If it is used in the integrated mode, the initialization step which loads the necessary data files from the hard drive can be preponed so it has to be run only once.

4.1.2. Atmospheric Model

The atmospheric module in FLORA uses data generated by the NRLMSISE-00 model described in section 2.1.1.2 to provide values for atmospheric density and scale height. It calculates the orbital perturbations using the analytical functions given by [King-Hele, 1987]. Details of the actual implementation are explained by [Radtke, 2011]: The data from the NRLMSISE model is given in the form of a lookup table that was generated by running the model about 800,000 times with all permutations of the input values given in table 4.1. For each of these value sets, the resulting atmospheric density and scale height are listed. The density values are averaged over the whole geographic latitude and one solar day. Since this introduces inaccuracies for highly eccentric orbits a correction function is applied later that eliminates the error in these cases. For each propagation step, the altitude is calculated from the object's current position. The other values are looked up in another table based on space weather data provided by [Kelso, 2000]. It contains measured mean and daily $F_{10.7}$ values as well as three-hourly A_p values for each day. For future dates, mean values are added assuming a medium solar activity. To obtain the highest possible accuracy from the atmospheric data table, a complex interpolation step is performed. For each of the four input values, the two closest matching indices within the range and step size given in table 4.1 are calculated. Values for density and scale height are looked up for every permutation of these indices resulting in a total of sixteen density/scale height pairs. These values are interpolated and returned. They are subsequently used as input for King-Hele's equations to determine the perturbed orbital elements. These equations require solving a modified Bessel function of the first kind for which a third lookup table is provided.

Input	Range	Step Sizes
$F_{10.7}$ (daily)	50 to 300	10
$F_{10.7}$ (81 day average)	50 to 290	15
A_p	0 to 175	15 (from 0 to 75) 25 (from 75 to 175)
Altitude (km)	100 to 4000	10 (from 100 to 1000 km) 20 (from 1000 to 2000 km) 50 (from 2000 to 4000 km)

Table 4.1.: Input values for which NRLMSISE data is provided in FLORA.

4.1.3. Third Body Perturbations

As described in section 2.1.1.3, five parameters are required to calculate the perturbations caused by a planetary body: Its gravitational parameter, called μ_3 in [Vallado, 2007], its equatorial inclination (i_3) and right ascension of the ascending node relative to the Earth (Ω_3), the argument of mean longitude (u_3) and the distance to the Earth (r_3). For the gravitational parameters of the two bodies the well-known constants are used. The equations for the other parameters as implemented in FLORA are given in [Flegel, 2007]. Most of them are calculated more accurately than the approximate constants given in literature based on the current Julian date in centuries. For example, the Sun's inclination, which is defined by the Earth's axial tilt, is given by [Vallado, 2007] as roughly 23.5°. In FLORA the value is calculated as

$$23.439291 - 0.0130042 \cdot T_{tdb} \tag{4.1}$$

where T_{tdb} is the current Julian century.

The third body perturbations in FLORA are implemented in three simple functions. Two collect the five parameters for the Sun and the Moon, respectively. The current propagation time is the only input required for these as all values are either constant or depend on T_{tdb} in the above manner. Once the parameters are collected they are given to a third function which uses them to calculate the third body's position relative to the satellite, and subsequently, the orbital perturbation according to the equations given by [Vallado, 2007].

4.1.4. Solar Radiation Pressure

Solar radiation pressure is implemented in FLORA based on the model described in section 2.1.1.4. It is assumed as constant; the acceleration it forces on the satellite is determined based on its reflectivity coefficient and area-to-mass ratio. Using this value as well as the Sun's position, the RSW coordinates describing the direction of the acceleration force are determined. Escobal's shadow function is used to check whether the object is currently within the influence of the Earth's shadow; if it is, the true anomalies of entry and exit points are calculated as described in section 2.1.1.4. Based on the outcome, the appropriate set of analytic equations from [Vallado, 2007] is chosen to determine the orbital perturbation.

4.1.5. Zonal Harmonics

The gravitational perturbations used in FLORA are an implementation of the analytical model given in [Vallado, 2007]: Equations 9-38, 9-40 and 9-42 describe the changes of the right ascension of the ascending node, argument of perigee and mean anomaly, respectively, as a function of time. FLORA supports zonal harmonics terms J_2, J_4 and J_6; sectoral and tesseral harmonics are not supported. In the implementation, the equations are divided so that the sections corresponding to the respective zonal terms can be calculated separately. This allows the user to configure the desired accuracy.

4.1.6. FLORA as an OPI Plugin

FLORA has been updated constantly over the years. One of the more recent changes included structural redesigns which separated the frontend (consisting of file input/output and configuration subroutines) from the actual propagation. This made adding the OPI interface a very simple task. It was first suggested in [Möckel et al., 2012] and finally described in [Köhncke, 2014], a thesis in which the OPI interface was used to automatically compare different propagators: The implementation involves a simple wrapper that lets the OPI function *OPI_Plugin_propagate* call the original FLORA propagation routine shown in listing 4.1 and passes on the arguments collected from *OPI::Population*. After successful execution of the propagation function, the output is simply written back to the population and the function *OPI_Population_update* is called to register the changes. To mark objects that have reentered into the atmosphere in the given propagation step, FLORA uses a "reentered" flag that is set to a negative value in case of reentry. This value is evaluated by the wrapper function; in case of a detected reentry, the EOL value of the respective OPI::Orbit is set to the given Julian day. The wrapper also contains implementations of the functions *OPI_Plugin_Info* for general information on the plugins and *OPI_Plugin_init* for registering properties for configuration. The properties contain user-definable settings on which perturbation forces should be considered; they are simply mapped to the respective global variables used in FLORA.

Since FLORA only supports the propagation of a single object, the wrapper loops over all objects of the *OPI::Population* and calls the propagation function separately for each of them. With the wrapper in place, creating an OPI plugin from FLORA is merely a matter of adding the *-shared* flag to the Fortran compiler to create a shared library instead of a standalone executable. The program will then be recognized by OPI as a propagator plugin and can be used by any host application.

Listing 4.1: Original FLORA subroutine used to execute a propagation.

```
SUBROUTINE floralib(
&       jd_init,       ! --> epoch / julian date [-]
&       time,          ! --> propagation time frame [d]
&       dt,            ! --> propagation step size[s]
&       el,            ! <-> orbit elements
&       Am,            ! --> area-to-mass ratio [m^2/kg]
&       CD,            ! --> drag coefficient [-]
&       CR,            ! --> reflectivity coefficient [-]
&       reenter,       ! <-- reentered flag
&       data_dir,      ! --> directory of input files
&       output_dir,    ! --> directory of output files
&       ich_flora,     ! --> channel of...
                       ! (0) .inp file
                       ! (1) expatm.dat file
                       ! (2) modbessel.dat file
                       ! (3) sw19571001.txt
                       ! (4) output file (0 = no output)
&       init_flag      ! --> .FALSE. for initialization
)                      !     .TRUE. for propagation
```

4.2. Ikebana - A Parallel CUDA Propagator

4.2.1. Overview

Ikebana was designed and implemented to work as an OPI plugin. Not only does it use the OPI interface to communicate with its host, it also follows the structural design that is suggested by the library. OPI's design is also reflected in the modularization which separates control code, calculation of an object's movement on its orbit, and physical models for perturbation forces. The latter are implemented as C++ classes which share the common interface defined in *OPI::PerturbationModule*. This interface forces the perturbation models to work with OPI's own data structures which are laid out as arrays instead of single objects, thus allowing them to be designed for arbitrary population sizes. It also allows separating time-dependent sections of the algorithm from those that are object-dependent. Both are significant changes from FLORA's structure which allow Ikebana to be optimized for propagation of multiple objects. All classes in Ikebana are tightly integrated to work as dedicated functional units. The use of global variables is avoided as it is often a source of confusion, especially in an environment that features asynchronous parallel execution of multiple threads. Besides the perturbation modules, Ikebana consists of a main class containing control code and the OPI interface to the host, and an auxiliary class that manages the precalculated NRLMSISE data. Each component is implemented in C++ with embedded CUDA kernels containing the parallelized calculations. Ikebana uses OPI's data types throughout its entire code, thus avoiding unnecessary conversion. All CUDA kernels run at a block size of 256 threads per block and a grid size depending on the number of objects in the population. Only the atmospheric model uses optimized block and grid size settings (see section 5.3.5).

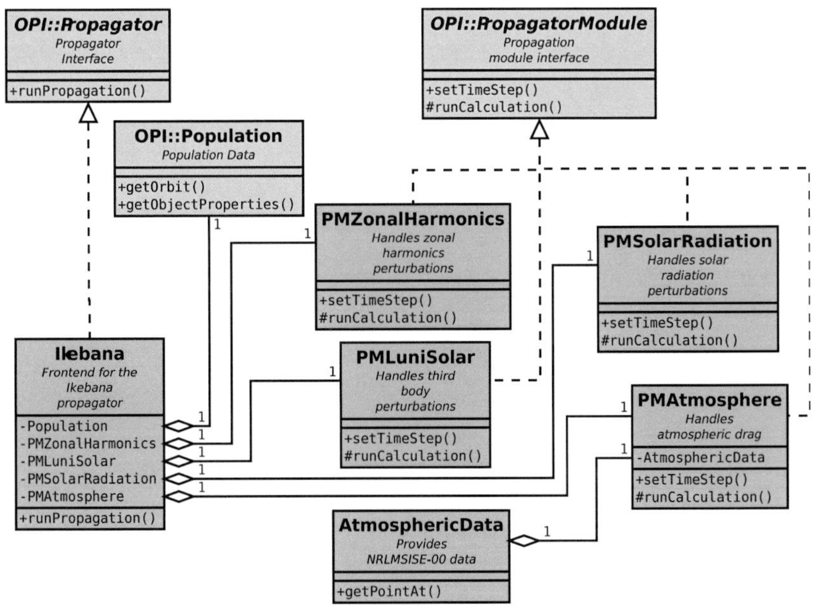

Figure 4.1.: UML diagram of Ikebana.

Figure 4.1 shows the main classes of Ikebana which are described in the following sections. For each of the more complex classes a flow chart is shown in a separate figure (4.2, 4.3, 4.4 and 4.7, respectively) depicting the basic steps through the algorithm. Serial parts that are executed on the CPU are drawn in green, everything that happens in a CUDA kernel is highlighted in blue. Input and output data are shown in yellow. The header files containing the complete class definitions, including function names that are referenced in this chapter, can be found in appendix C. Auxiliary classes that are not explicitly mentioned are *AstroMath* which contains generic mathematical functions and astronomical constants, *AstroMathCUDA*, a parallelized port of this library, *IkebanaFileManager* which handles file input/output operations, *Voice* which manages the output of debug and logging information, and the Ikebana Frontend which serves as a minimal OPI host with the ability to read and write data in FLORA's native format, thus completing compatibility with FLORA.

4.2.2. Ikebana::Ikebana

The main class of Ikebana implements the *OPI::Propagator* interface and therefore functions as the connector between the host and the propagation. In the constructor, PropagatorProperties are set that allow the user to activate or deactivate individual perturbation models and the generation of Cartesian coordinates, set the J-terms to be accounted for in the zonal harmonics module and the amount of verbose output. The OPI functions *runEnable* and *runDis-*

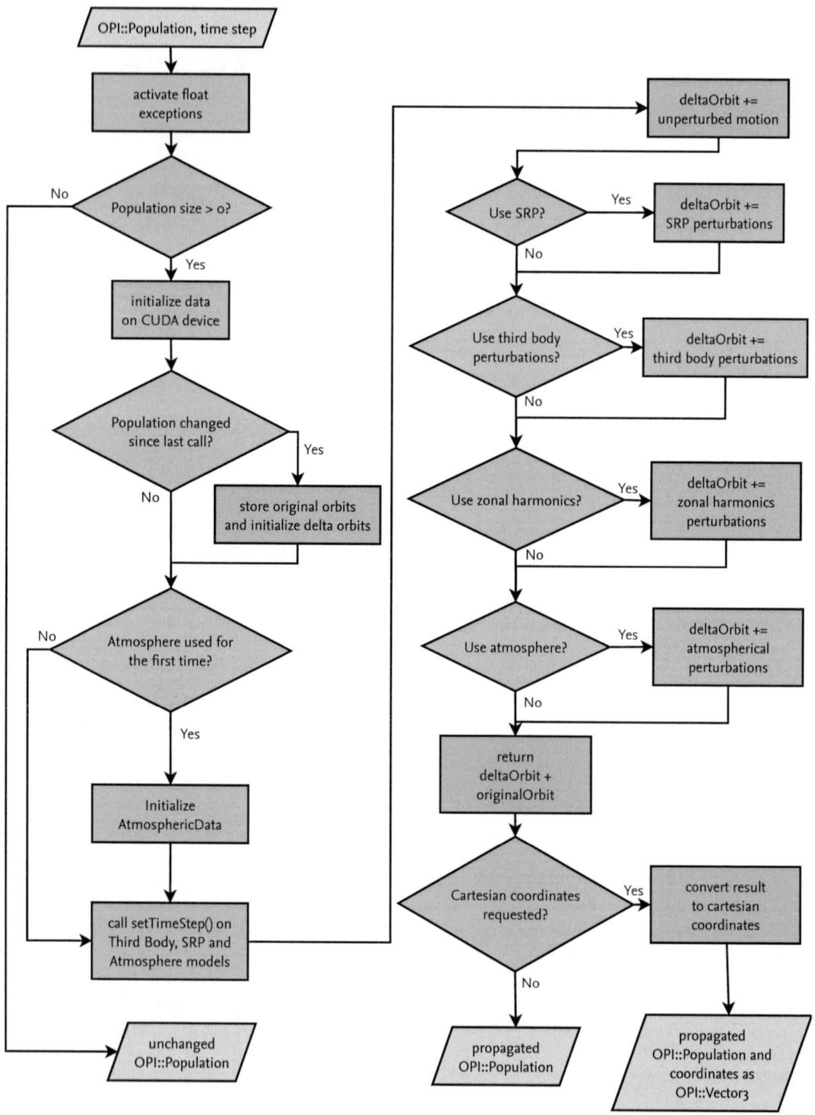

Figure 4.2.: Flowchart of the *runPropagation* function in Ikebana's main class.

able are overridden only to provide a message output for these events. The main functionality is implemented in the *runPropagation* function which is also inherited from OPI::Propagator. First of all, the function checks whether input data is available. If this is the case, pointers to the orbits and ObjectProperties of the given OPI::Population are initialized on the CUDA device. If necessary, OPI will transfer population data to the GPU memory automatically at this point.

Next, the function checks whether the size of the input data has changed since it was called last. Most notably, this happens on the first call. In this case, an initialization step is performed on the GPU. Device memory is allocated to store the original, unperturbed orbits as well as the *delta orbits* which are the sets of orbital elements representing the aggregated changes caused by the perturbation forces. The reason for storing delta orbits separately is to prevent loss of information due to floating point inaccuracies as explained in section 5.2.1. The delta orbits are set to zero; information from the host is checked for validity. Invalid values, such as undefined drag and reflectivity coefficients are set to the commonly used default values (according to [Vallado, 2007]) of 2.2 and 1.3, respectively. On the CPU, the lookup tables for the *AtmosphericData* class are loaded into memory if atmospheric perturbations are enabled.

After this initialization step, the actual propagation begins. First, the current time step is handed to the perturbation modules that require it via their respective *setTimeStep* functions. Each of them performs time-dependent precalculations on the CPU. The unperturbed motion is calculated on the GPU using the kernel described in section 4.2.3. Depending on which perturbation modules the user has enabled, those modules' respective *calculate* functions are called which run the CUDA kernels in which the perturbation models are implemented. These are described in detail in the following sections. The perturbed orbits they output are added to the delta orbit. Finally, in a post-processing step on the GPU, the delta orbit is added to the original orbit and written back to the *OPI::Population* instance. If Cartesian coordinates are requested, they are calculated from the new orbit on the GPU and written to the population as well.

The reason for checking the activation status of each module at each propagation step is to provide a high level of flexibility to the user. For the visualization use case in particular, it was found useful to start with just the mean motion activated by default since it provides a smooth frame rate even with a very high number of objects. When checking for module activation at every step, individual perturbations can be toggled by the user at any point during run time via a GUI element or keyboard command. Since the AtmosphericData class is initialized when it is first required, unnecessary load times and memory consumption are avoided.

In the *runPropagation* function, floating point exceptions are enabled for invalid results as well as overflows and underflows. This is to ensure that operations resulting in NaN or undefined values cause the program to stop with an error message. The reason for this is usually invalid input; as defined in the OPI guidelines, the host program should check the input data for correctness in these events.

4.2.3. Ikebana::PMMeanMotion

The perturbation module for the satellite's mean motion is the simplest since it only calculates a single equation and adds the result to the provided delta orbit:

$$M = \sqrt{\frac{\mu}{a^3}} \tag{4.2}$$

In Ikebana, this module is the minimum of what can be considered a propagation. Even with all perturbation forces and conversion to Cartesian coordinates switched off the mean motion is still calculated. Because of its simplicity it constitutes the smallest possible implementation of an *OPI::PerturbationModule* using CUDA and can be used as an easy-to-understand example. The module's definition is shown in listing 4.2. After the "include guard" and inclusion of the OPI header, the class *Ikebana::PMMeanMotion* is defined as a derivative of *OPI::PerturbationModule*. From that class it inherits a number of functions, two of which have to be overridden in order to implement the module's functionality: The public *setTimeStep* and the protected *runCalculation*. Because of current API limitations, CUDA functions have to be defined outside of classes; in this case, one global CUDA function named *meanMotion* exists which does the actual calculation on the GPU.

Listing 4.2: *PerturbationModule* implementing normal unperturbed motion (header).

```
#ifndef __PERTURBATION_MEAN_MOTION_H__
#define __PERTURBATION_MEAN_MOTION_H__

#include "OPI/opi_cpp.h"

class PMMeanMotion: public OPI::PerturbationModule
{
  public:
    OPI::ErrorCode setTimeStep(double julian_date);

  protected:
    OPI::ErrorCode runCalculation(OPI::Population& data, OPI::Orbit* delta, float dt);
};

__global__ void meanMotion(OPI::Orbit* orbit, OPI::Orbit* delta, int size, float dt);

#endif
```

The class' implementation is shown in listing 4.3. The included header files contain the class definition shown above as well as basic mathematical functions, constants such as μ and the default CUDA grid and block sizes. The function *setTimeStep* does nothing since the mean motion is independent of the date. *runCalculation* takes as argument the current population, an equally-sized array of delta orbits and the current step size. It calls the CUDA kernel *meanMotion*, passing on the information to the GPU. The variable *idx* contains the objects' individual indices; since OPI lays out the population data one-dimensionally, all kernels working on an *OPI::Population* must calculate the indices in the same way. Using the index, each thread processes equation 4.2 for the population object assigned to it and adds the result to the mean anomaly of the delta orbit with the corresponding index. No double precision values are used within this module.

Listing 4.3: *PerturbationModule* implementing normal unperturbed motion (implementation).

```
#include "PMMeanMotion.h"
#include "AstroMathCUDA.h"
#include "config.h"

OPI::ErrorCode PMMeanMotion::setTimeStep(double julian_date)
{
  return OPI::SUCCESS;
}

OPI::ErrorCode PMMeanMotion::runCalculation(
  OPI::Population& data, OPI::Orbit* delta, float dt)
{
  calculateMotion<<<GRID_SIZE, BLOCK_SIZE>>>(
    data.getOrbit(OPI::DEVICE_CUDA, true), delta, data.getSize(), dt
  );
  return OPI::SUCCESS;
}

__global__ void meanMotion(OPI::Orbit* orbit, OPI::Orbit* delta, int size, float dt)
{
  int idx = blockIdx.x*blockDim.x + threadIdx.x;
  if (idx < size) {
      const float meanMotion =
      sqrt(ASTROMATH_EARTH_GRAVITATION / pow(orbit[idx].semi_major_axis,3.0f));
    delta[idx].mean_anomaly += AstroMath_moduloRad(meanMotion * dt);
  }
}
```

4.2.4. Ikebana::PMZonalHarmonics

Like in FLORA, the CUDA kernel of the zonal harmonics perturbations module is a straightforward implementation of the analytical equations 9-38, 9-40 and 9-42 from [Vallado, 2007] that describe the changes of the right ascension of the ascending node, argument of perigee and mean anomaly as a function of time. The implementation is equivalent to FLORA's which supports zonal harmonics terms J_2, J_4 and J_6. The only difference lies in the use of single precision variables throughout the module in Ikebana where FLORA uses double precision. Besides the objects' respective orbits and five constants - the Earth's gravitational parameter, its mean radius and the three supported J-terms - no additional input data is required. Since there are no time-dependent operations in this module, the *setTimeStep* function inherited from *OPI::PerturbationModule* does nothing. The *runCalculation* function is also very simple and merely executes the CUDA kernel. Like in FLORA, the equations are divided so that the sections corresponding to the respective zonal terms can be calculated separately. Only minor changes have been made such as moving some definitions of constants and variables to resolve scoping issues. The module implements its own PropagatorProperty to allow the J-term accuracy to be configured by the user. In Ikebana, the evaluation of this setting is handled by the main class and transferred to the zonal harmonics module at each call. The choice to implement it as a PropagatorProperty instead of a class variable has been made to facilitate detaching the class into an individual module. In terms of computational complexity, reducing accuracy by skipping J_4 or J_6 terms has virtually no effect on the overall runtime.

The kernel is shown in listing 4.4. Like all of the more complex perturbation modules in Ikebana, it first checks whether the object has already decayed by querying its end of life value. If it's still in orbit, the actual calculation begins. First some general constants are defined;

after that, the changes in RAAN, argument of perigee and mean anomaly are added to the delta orbit for the J_2 term using the respective portions of Vallado's equations. Depending on the setting of the *PropagatorProperty* defining the J-terms to be used, the same is done for J_4 and J_6. Since the equation give the change in radians per second, the results are multiplied with the current step size, dt. Finally, some additional settings are performed that all perturbation modules need to do to ensure consistency. First of all, it is shown that a temporary variable was used instead of changing the delta orbit directly. While the latter wouldn't pose a problem in this case, sometimes previous values from the delta orbit are required in following equations; a temporary variable must be used to prevent them from being overwritten. All unused variables of the temporary orbit must be initialized to zero; otherwise, undefined values might get added to the delta which may change the results in unexpected ways. Lastly, if the argument of perigee was changed by the perturbation force it needs to be subtracted from the mean anomaly. The reason for this is that the mean anomaly calculated by *Ikebana::PMMeanMotion* is based on the original orbit. Since the mean anomaly is offset from the argument of perigee by definition, changing this value later would falsify the satellite's position. This is prevented by subtracting the change from the mean anomaly.

Listing 4.4: Excerpt from the kernel that calculates gravitational perturbations in Ikebana showing the structural design of the function.

```
__device__ void gravity(OPI::Orbit& orbit, OPI::Orbit& deltaOrbit, float dt, int jTerm)
{
  if (orbit.eol > 1.0) {
    // object already decayed, do nothing
  }
  else {
    // some important constants and simplifications
    const float meanMotion = sqrt(EARTH_GRAVITATION/pow(orbit.semi_major_axis,3));
    const float parameter = orbit.semi_major_axis * (1 - pow(orbit.eccentricity,2));
    const float rp2 = pow(ASTROMATH_EARTH_RADIUS,2) / pow(parameter,2);
    const float e2 = pow(orbit.eccentricity,2);
    const float sin2i = pow(sin(orbit.inclination),2);

    OPI::Orbit tmp; // temporary variable for the delta orbit

    if (jTermAccuracy >= 2) {
      // J2 constant and abbreviations specific to J2 term
      const float j2Term = 1.08262668355315e-03f;
      const float mmj2 = meanMotion * j2Term * rp2;
      const float mmj2sq = mmj2 * j2Term * rp2;

      // calculate J2 RAAN - Vallado (9-38) in [rad/sec] times dt
      tmp.raan = dt *
        ((-1.5f * mmj2 * cos(orbit.inclination)) +
        (3.0f * mmj2sq  * cos(orbit.inclination) / 32.0f) *
        (12.0f - 4.0f * e2 - ((80.0f + 5.0f * e2) * sin2i)));

      // same for J2 argument of perigee (9-40) and mean anomaly (9-42)
    }

    if (jTermAccuracy >= 4) { /* same for J4 terms */ }
    if (jTermAccuracy >= 6) { /* same for J6 terms */ }

    // remove change in argument of perigee from mean anomaly to prevent it from
    // being counted twice
    tmp.mean_anomaly -= tmp.arg_of_perigee;

    // set unused variables to zero
    tmp.eccentricity = 0.0f;
    tmp.semi_major_axis = 0.0f;
    tmp.inclination = 0.0f;
    tmp.bol = 0.0f;
    tmp.eol = 0.0f;

    // add to delta orbit
    deltaOrbit = deltaOrbit + tmp;
  }
}
```

4.2.5. Ikebana::PMLuniSolar

Ikebana's default module for luni-solar perturbations is very similar to the original implementation in FLORA, except that all variables have been converted to single precision. The luni-solar module is particularly suitable for demonstrating the usefulness of OPI's distinction between setting the time step and performing the propagation. A great deal of the calculations are time-dependent but not orbit-dependent and can therefore be precalculated on the CPU before the CUDA kernel is executed. This is shown in figure 4.3.

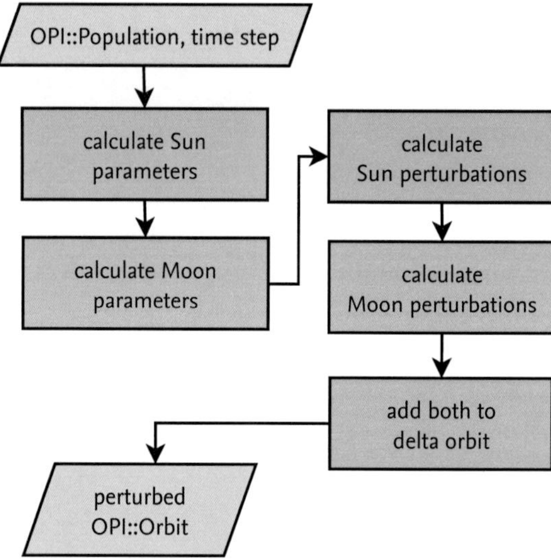

Figure 4.3.: Flowchart of Ikebana's luni-solar PerturbationModule.

Since the five parameters, μ_3, i_3, r_3, u_3 and Ω_3 are equally required for both planetary bodies they are combined into a struct called *tThirdBody*. Like in FLORA, they are collected by two functions; their equivalents in Ikebana are called *getSunParameters* and *getMoonParameters*. These functions are called by the *setTimeStep* function that take as input the current Julian date and returns an instance of *tThirdBody* filled with the parameters for the respective body. They are stored as class attributes so the *calculate* function can access and upload them to the GPU later.

Once all parameters for Sun and Moon are determined for the current time step, perturbances can be calculated for each object. When the *calculate* function is called by Ikebana's main class it transfers the third body parameters to the GPU via a CUDA kernel call. The CUDA kernel first checks for each object whether it has already decayed. For those which are still in orbit, the CUDA function *calculateThirdBody* is called twice - first with the Sun parameters, then with the Moon parameters. The function uses the orbital and third body data to calculate the direction cosines A, B and C and subsequently the perturbations of the orbital parameters according to [Vallado, 2007], equation 9-54. Since the function calculates the changes per second, the results from both calls are multiplied with the current step size and added to the delta orbit.

4.2.6. Ikebana::PMSolarRadiation

This *PerturbationModule* is responsible for the calculation of the solar radiation pressure perturbation in Ikebana. The flowchart is shown in figure 4.4. The implementation from FLORA received only structural updates. The majority of the algorithm is located in the CUDA kernels because it depends on individual orbital elements as well as reflectivity coefficient and

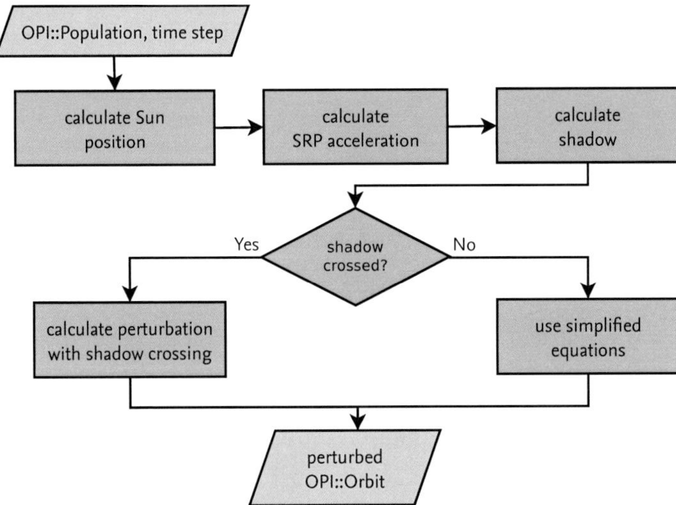

Figure 4.4.: Flowchart of Ikebana's solar radiation pressure PerturbationModule.

area-to-mass ratio from *OPI::ObjectProperties*. The only independent value is the Sun's position whose calculation is called from the *setTimeStep* function. Two intermediate results from this function, the exact obliquity of the Earth at the given date as well as the Sun's ecliptic longitude, are stored for upload to the GPU since they are required by the kernel for subsequent operations.

The kernel which is called by *runPropagation* first calls the device function *getSRPAcceleration* which calculates the acceleration of the solar radiation pressure as described in section 2.1.1.4. The four components, R_P, S_P, $W_{\sin(\omega)}$ and $W_{\cos(\omega)}$, are returned as a struct called *tAcceleration*. The next step is carried out in the aptly named *calculateShadow* function. First, the perifocal coordinates β and ξ are calculated which, together with the orbit's semi major axis and the eccentricity, serve as input to solve the quartic equations given by [Escobal, 1965]. Contrary to the rest of the kernel, the quartic equations are solved using double precision variables because it was found that single precision floats would underflow in some cases. The results are used to determine the entry and exit points of the satellite via Escobal's shadow function. Another struct called *tShadow* is returned that contains a boolean variable that states whether the shadow was crossed in this propagation step; it also contains the eccentric and mean anomalies as well as the radius vectors of the shadow entry and exit points. Although these values are not required when the shadow was not crossed, they are calculated in both cases. Since the threads that do not need this information have to wait for others to finish, this will affect the run time only in the very unlikely situation that all objects in one block are outside the shadow. Finally, with all necessary values gathered the function *calculateShadowInfluence* is called which generates the orbital perturbations based on Vallado's equations; depending on the outcome of the shadow crossing test, the appropriate set is chosen.

4.2.7. Ikebana::AtmosphericData

The AtmosphericData class is the only major class used in Ikebana that is not derived from OPI. It is used as a data manager for *Ikebana::PMAtmosphere* and provides access to the lookup tables containing the NRLMSISE-00 data and solar activity. These tables are identical to the ones in FLORA. A third lookup table is provided for the required modified Bessel function of the first kind; CUDA provides a number of Bessel functions as part of its math libraries but since the one required is not among them the lookup table approach from FLORA was adopted. FLORA performs an interpolation step on this table upon loading; to save some initialization time, Ikebana omits this step by using a different table created from a memory dump of FLORA's interpolated values. For convenience and to save disk space, all tables are packed into a zip archive called *ikebana.dat* and loaded directly from there via the PhysicsFS library ([Gordon, 2010]) which provides an abstraction layer for file system operations. The tables are loaded automatically when they are required for the first time and kept in memory until Ikebana is disabled.

Like in FLORA, Ikebana's atmospheric model requires data for density and scale height. These values are looked up and interpolated based on the four required inputs: Altitude, mean and daily $F_{10.7}$ values and the day-average A_p (figure 4.5). To obtain the maximum efficiency in the parallel portion of the code, Ikebana handles this step differently from FLORA. Of the four required input values, only the altitude depends on an object's individual property; the others only depend on the date and do not have to be determined by the kernel. On initialization of the atmospheric module, the lookup table containing solar activity data is loaded into CPU memory while the atmospheric data table is copied to the GPUs global memory. In addition, three unified memory integer pointers are created for the A_p and the two $F_{10.7}$ values. During propagation, upon calling *setTimeStep* on *Ikebana::PMAtmosphere*, the three numbers are looked up based on the given Julian date and stored in unified memory. During propagation, each thread can now use these values together with the individual object's current altitude as indices into the atmospheric data table to read the correct density and scale height information (figure 4.6).

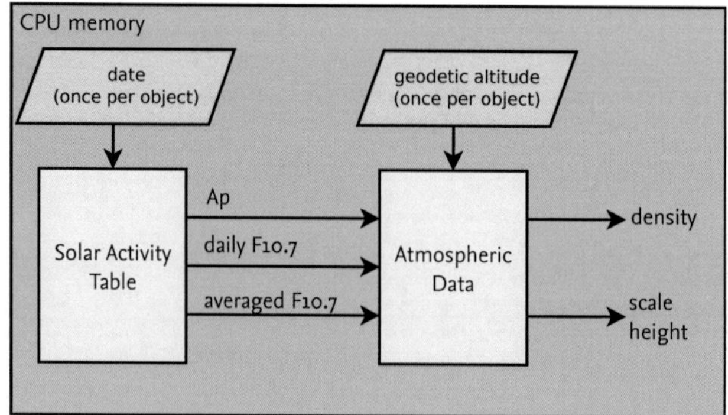

Figure 4.5.: Handling of atmospheric data in FLORA.

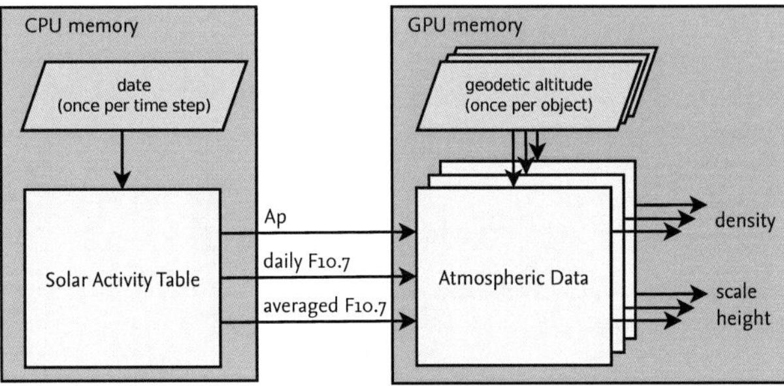

Figure 4.6.: Handling of atmospheric data in Ikebana.

4.2.8. Ikebana::PMAtmosphere

Ikebana's atmospherical model uses *Ikebana::AtmosphericData* for accessing lookup table data. An instance of it is maintained by *Ikebana::PMAtmosphere* which serves as the main class for this PerturbationModule. Figure 4.7 shows the combined flowchart for the two. On the module's first use, the *PMAtmosphere* class calls the initialization routine of the data provider to load the data from the lookup tables into their respective memory areas. As explained above, the *setTimeStep* function calculates the indices for three of the four input values based on the current Julian date and makes them accessible to the kernel. When the actual kernel is run by calling the *runPropagation* function, first the geodetic altitude of each object is calculated. If it is larger than 4000 km, the object is assumed to be outside the atmospheric influence and is returned unaltered. Otherwise, the geodetic altitude serves as the fourth and final index into the atmospheric data table. Since this was stored in GPU memory by the data provider, the kernel can now aquire the correct density and scale height for each object. This is done via the data provider's *getDataAt* function which takes as input the geodetic altitude as well as pointers to the atmospheric data table and the previously calculated indices for A_p and $F_{10.7}$ values. The points around these indices are chosen and interpolated with functions equivalent to FLORA's as described in section 4.1.2. Next, the ballistic coefficient is calculated from the object properties given in the *OPI::Population*. With all required input values collected, King-Hele's equations can now be solved; depending on the orbit's eccentricity, the corresponding set is chosen. The Bessel function required for solving these equations is calculated in double precision because some resulting values would overflow a single-precision float. The determination of the change in eccentricity is also conducted in double precision because it was found to cause inaccuracies for small values otherwise.

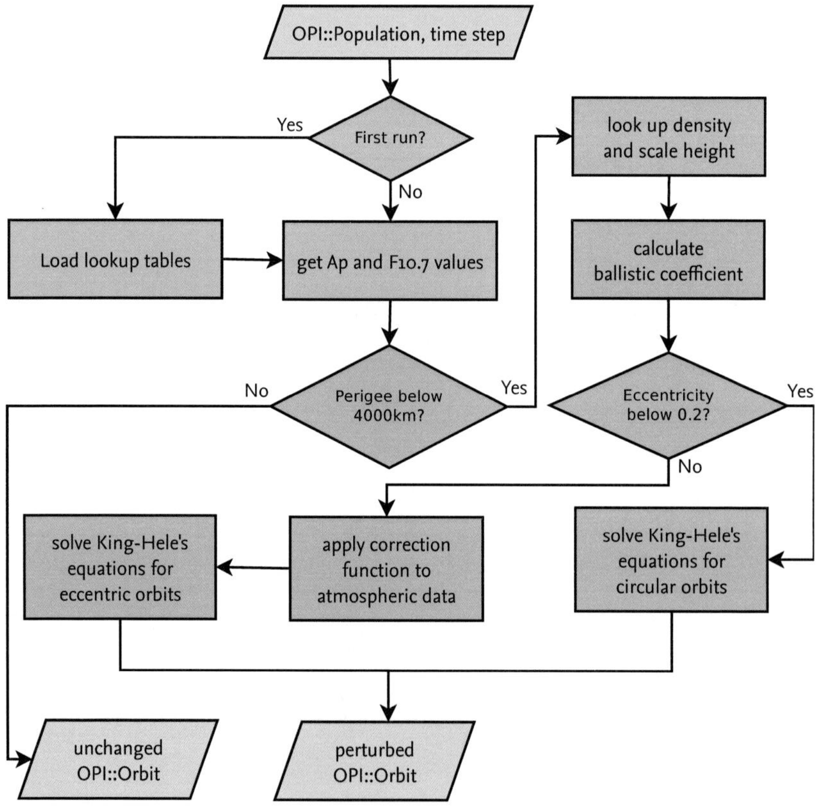

Figure 4.7.: Flowchart of Ikebana's Atmospheric PerturbationModule (including functions from AtmosphericData).

5 Performance Analysis

5.1. Reference Population

For the task of comparing speed and accuracy between FLORA and Ikebana, a reference population was created based on publicly available data. A snapshot of the full TLE catalogue from [Space-Track.org, 2015] was used as the basis and transformed to single mean Kepler orbits with the SGP4 algorithm. Some missing information was added randomly, with value ranges representing realistic satellites:

- Mass between 500 and 5,000 kg,
- mean diameter between 1 and 9 metres,
- mass-to-area ratio between 40 and 480 kg per square metre,
- mean anomaly between 0 and 359 degrees.

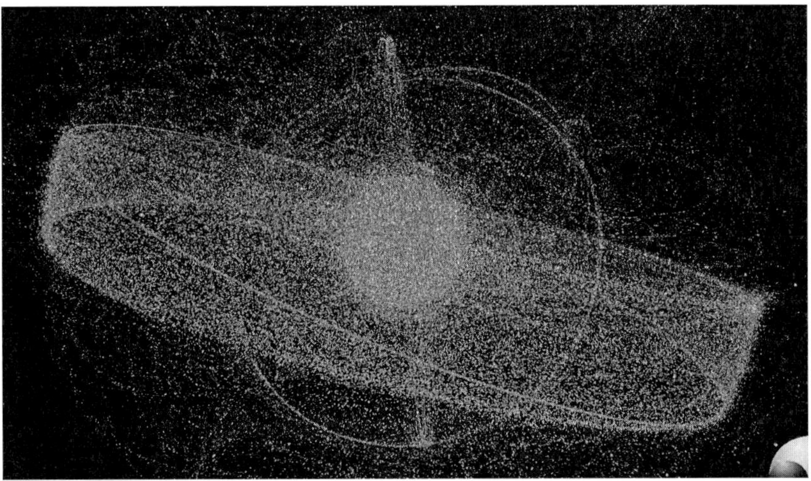

Figure 5.1.: The reference population used for propagator validation and analysis, visualized with DOCTOR.

Since the TLE snapshot contained only around 15,000 objects, this process was repeated 100 times to generate a catalogue of roughly 1.5 million objects. The whole resulting file was shuffled to ensure equal distribution of orbit types in case a smaller number of objects was picked.

5.2. Accuracy

5.2.1. Floating Point Considerations

Ikebana uses single precision floating point variables where possible because of their higher performance on the GPU ([NVIDIA Corporation, 2013]; also see section 5.4). Since FLORA was written with double precision throughout, the question arises whether this change will have an impact on the results. In [Fraire et al., 2013], for example, it is shown that the GPU port of SGP4 struggles with this issue: The switch from double to single precision causes errors that increase over time. The paper shows an average deviation in the object's position of about 12 km after approximately 23 years. Since the SGP4 algorithm is quoted as having a relative error margin of 1-3 km per day, this additional error is insignificant in the given context. However, it is still significantly larger than that of the double precision implementation which stays below 1 km and does not increase. Even though this example proves that reducing precision is not always possible without significant loss of information, the general use of double precision should be avoided for performance reasons when designing an algorithm for the GPU. To achieve the best possible tradeoff between speed and accuracy, the algorithm has to be studied carefully to identify the locations where single precision is sufficient, those where the use of double precision can be worked around by rearranging the equations, and those where double precision is unavoidably required. Some of these have already been mentioned in the previous chapter.

Floating point inaccuracies can occur when two numbers with different magnitudes are added. Consider the example shown in listing 5.1: The semi major axis of a GEO orbit, $42,131.114$ km, is stored to a single precision float variable. A perturbation of one metre is added to that value for ten consecutive time steps. In the first method, both numbers are added directly, resulting in the output shown in the leftmost column of table 5.1: The first issue that stands out is that the exact initial value for the semi major axis cannot be represented in single precision so it is rounded to the nearest representable value, $42,131.113281$. Applied to reality, this means that just by reading the value into a single precision float, the semi major axis is already off by roughly 72 centimetres. Secondly, the perturbation added to it is so small by comparision that it is annihilated in the rounding process. Every consecutive addition follows the same process, thus resulting in complete loss of the perturbation information. In other words, every time step adds one metre to the error.

To minimize this effect, Ikebana uses the second method shown in the algorithm: Instead of adding the perturbation to the semi major axis directly, a delta value is introduced that "collects" the changes by summing up only the perturbations, thus making sure that the summands stay in roughly the same magnitude. The final result is created by adding the collected delta to the original semi major axis. Since the delta value becomes larger after a few time steps, the rounding error is less severe. The second column of table 5.1 shows that while some information is still lost, the effect is much less destructive and, most importantly, does not carry into consecutive operations. The error stays around one metre during all time steps. However, if the number of steps becomes large enough for the delta value to enter the magnitude range of the original semi major axis, it should be added to it and reset to avoid the problem that method 1 demonstrates. Both methods, rewritten with double precision variables as shown in the rightmost column, are able to exactly represent the chosen values.

Listing 5.1: Example of precision loss when using floats and how to minimize it.

```
#include <stdio.h>

int main (int argc, char** argv)
{
  float semi_major_axis = 42131.114f;
  float perturbation = 0.001f;

  // Method 1:
  float result = semi_major_axis;
  for (int t=0; t<10; t++) {
    result += perturbation;
    printf("Method␣1,␣step␣%d:␣%f\n",t,result);
  }

  // Method 2:
  float delta = 0.0f;
  for (int t=0; t<10; t++) {
    delta += perturbation;
    result = semi_major_axis + delta;
    printf("Method␣2,␣step␣%d:␣%f\n",t,result);
  }

  return 0;
}
```

	Float, Method 1	**Float, Method 2**	**Double, Methods 1&2**
Initial value	42131.113281	42131.113281	42131.114000
t = 0	42131.113281	42131.113281	42131.115000
1	42131.113281	42131.117188	42131.116000
2	42131.113281	42131.117188	42131.117000
3	42131.113281	42131.117188	42131.118000
4	42131.113281	42131.117188	42131.119000
5	42131.113281	42131.121094	42131.120000
6	42131.113281	42131.121094	42131.121000
7	42131.113281	42131.121094	42131.122000
8	42131.113281	42131.121094	42131.123000
9	42131.113281	42131.125000	42131.124000

Table 5.1.: Results of the algorithm from listing 5.1.

The same problem occurs when the Julian date is represented by a single precision variable: Adding a small amount of time such as a single second to a recent Julian date will most likely result in that second being lost in the rounding process. It is possible to avoid this problem in a similar manner, for example by using two floats: One to count the years that have passed and one to count the seconds of the current year. However, the Julian date is only required in kernel code on two occasions in Ikebana: Once to calculate the correction factors for the atmospheric lookup table, and in the post-processing step to set the end-of-life date for decaying objects. All other values relying on the date are precalculated on the CPU and the results are uploaded to the GPU. Therefore, using a double precision variable for the date has no measurable effect on the GPU execution speed.

In the implementation of Ikebana with OPI, this solution has one major drawback: When Ikebana is initialized, an array of delta orbits is created with the same size of the input population (see section 4.2.2 for details). This array is managed by the propagator. When the host program changes the size or the order of population objects, the delta orbits do not match anymore; the propagation process up to that point is lost. In order to preserve this information, the host application has to download the current propagation process from the plugin, apply it to the updated population as required and then send the updated population back to the plugin. This is another example that shows the amount of implementation effort that is added by constraining the application to single precision floats.

5.2.1.1. Overflows, Underflows and NaN

Floating point overflows and underflows occur when operations result in a number that exceeds the largest and smallest representable value, respectively. In Ikebana, an example of a possible underflow can be found in the solar radiation pressure module when a very small eccentricity is raised to the power of four; the result may be too small to be represented in single precision and must be either stored in a double variable or raised artificially. NaN is a value defined in the floating point standard which is used to represent undefined values such as the result of a division by zero or the sqare root of a negative - the floating point type does not include complex numbers. Occurrences of overflows, underflows and NaNs in analytical propagation should be treated as programming errors. When using single precision for critical values like eccentricity, this might limit the domain of possible inputs which requires appropriate handling and documentation. Ikebana was tested with all objects of the reference population and found to produce no problematic output. In the event of overflows, underflows or NaN results, Ikebana terminates with an exception. This behaviour can be disabled in debug mode in order to track the problem.

5.2.2. Accuracy Dertermination

To test the accuracy of Ikebana, 1000 randomly chosen objects of the reference population were propagated over 50 years with both FLORA and Ikebana. As a starting date, January 1st, 1971, was selected arbitrarily, the time step size was set to one day. Both propagators were addressed through the OPI interface to ensure identical handling of input and output values. The software used for this is based on a work by [Köhncke, 2014] which demonstrates the usefulness of OPI for automated propagator validation. FLORA stores all intermediate results in double precision internally; small rounding errors may occur when fetching them through OPI, but consistent errors or loss of information can be ruled out. The original version of FLORA employs a different rounding process when writing the output files which causes tiny differences. Figure 5.2 shows a typical test case with all perturbations applied and negligible deviation. Propagations were conducted individually for every perturbation module; in a final scenario, all perturbation forces were activated. For all tests, representative individual object results are given that show average values or significant exceptions.

Many publications comparing orbital propagators cite the objects' positional errors as a benchmark for accuracy. However, for statistical long-term analysis of large populations, the main use case of Ikebana, individual objects' positions have little meaning; the shape and position of the orbit as a whole is more significant. Therefore, results are collected and displayed individually for semi major axis, eccentricity, inclination, perigee height and right ascension of the ascending node. Results for the argument of perigee are shown only for the gravitational perturbations. The mean anomaly values, representing the objects' positions, are omitted for the reason stated above.

It is important to note that the goal of the tests conducted in this section is to prove that analytical propagation can be carried out on the GPU despite the constraints that this platform poses in terms of floating point accuracy. It shall be proven that the deviations between FLORA and Ikebana are within an acceptable range for their designated use cases. Therefore, Ikebana aims to recreate FLORA's output as closely as possible, not to improve its absolute quality. In section 5.2.4, the errors caused by Ikebana's lower accuracy are compared against errors introduced by uncertainties in other input values such as solar activity and atmospheric density; inaccuracies within this magnitude are considered acceptable. An example of the propagators' performance against real measurement data is shown in section 5.2.5.

5.2.3. Individual Model Accuracy

5.2.3.1. Gravitational Perturbations

Ikebana's gravitational perturbations module shows no significant deviations from FLORA. The changes in RAAN and argument of perigee match FLORA's output very well in all orbital regimes. Differences are generally below 10 degrees which is acceptable since this value changes at a high rate. The overall progress of the curve is the same in both propagators. Figures 5.3 and 5.4 show examples for LEO and GEO orbits, respectively. The other elements (with the exception of the mean anomaly) are not affected.

5.2.3.2. Third Body Perturbations

The third body perturbations module shows very good overall results in all orbital configurations. The vast majority of objects display no significant errors as the examples show (figures 5.5 and 5.6). In a few cases, a slight shift in the eccentricity of GEO orbits was observed, causing the orbit height to deviate (figure 5.7). The error is introduced at a specific point in time and stays relatively constant afterwards.

Listing 5.2: The lines of MPLuniSolar that cause the inaccuracy visible in figure 5.7.

```
float julianCentury = (julianDate - 2451545.0f) / 36525.0f;

const double moonRAANEcliptic = AstroMath_moduloDeg(
  125.04455501f - (5.0f * 360.0f + 134.1361851f) * julianCentury
  + 0.0020756f * pow(julianCentury,2) + 2.136e-6f * pow(julianCentury,3)
  - 1.65e-8f * pow(julianCentury,4)) * ASTROMATH_DEG2RAD;

const double moonMeanAnomaly = AstroMath_moduloDeg(
  93.27209061f + 483202.0172f * julianCentury
  - 0.003542f * pow(julianCentury,2)
  - 2.88056e-7f * pow(julianCentury,3)
  + 1.15833e-9f * pow(julianCentury,4)) * ASTROMATH_DEG2RAD;
```

The problem is located in the two lines in *Ikebana::PMLuniSolar* that calculate the Moon's RAAN and mean anomaly, shown in listing 5.2. Floating point inaccuracies cause a slight error in the Moon's position and thus in the direction of its gravitational pull. Reverting the constants to double precision was able to noticably reduce the severity to the acceptable level shown in the figure. Since the code for determining the third bodies' positions runs entirely on the CPU the change did not affect the runtime speed. The remaining error is introduced when the position vector is converted to single precision on transfer to the GPU.

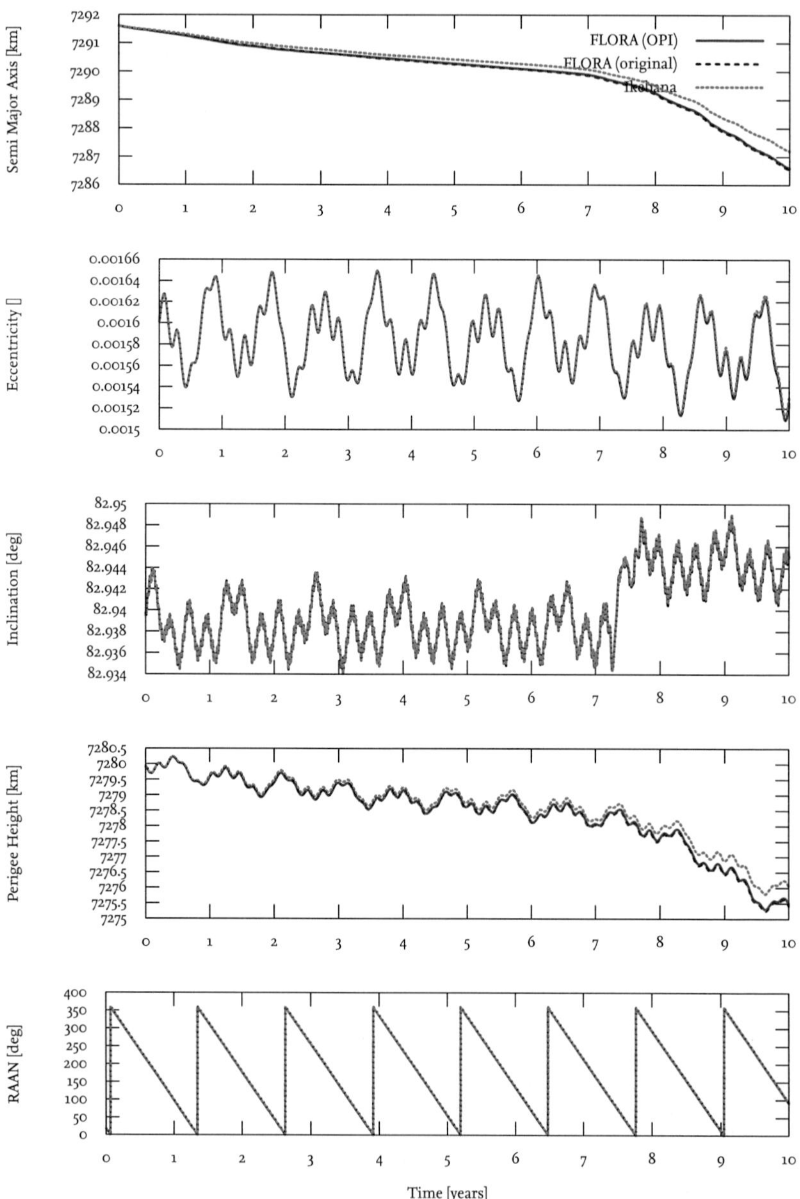

Figure 5.2.: Comparison between Ikebana and FLORA with and without OPI. Very small differences between the two FLORA versions are caused by slightly different rounding of the output data.

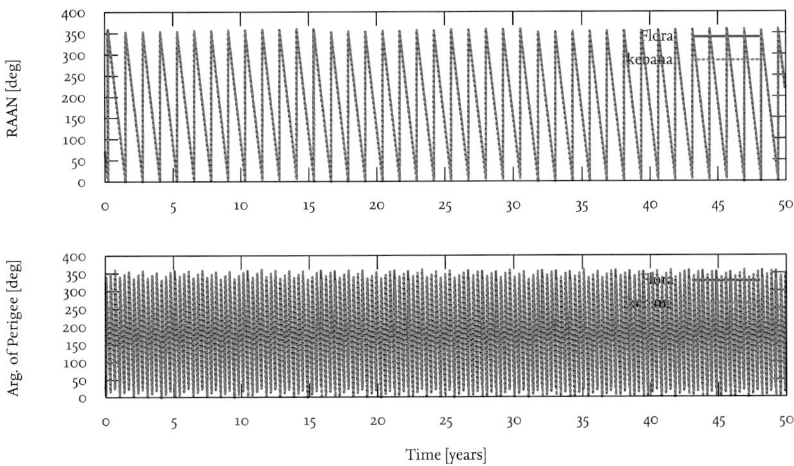

Figure 5.3.: Comparison of the zonal harmonics module of Ikebana and FLORA on a LEO orbit. Object number: 13464

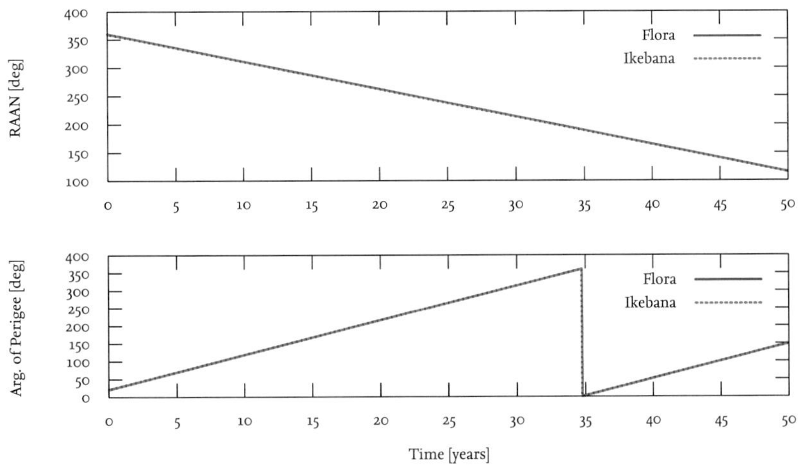

Figure 5.4.: Comparison of the zonal harmonics module of Ikebana and FLORA on a GEO orbit. Object number: 28472

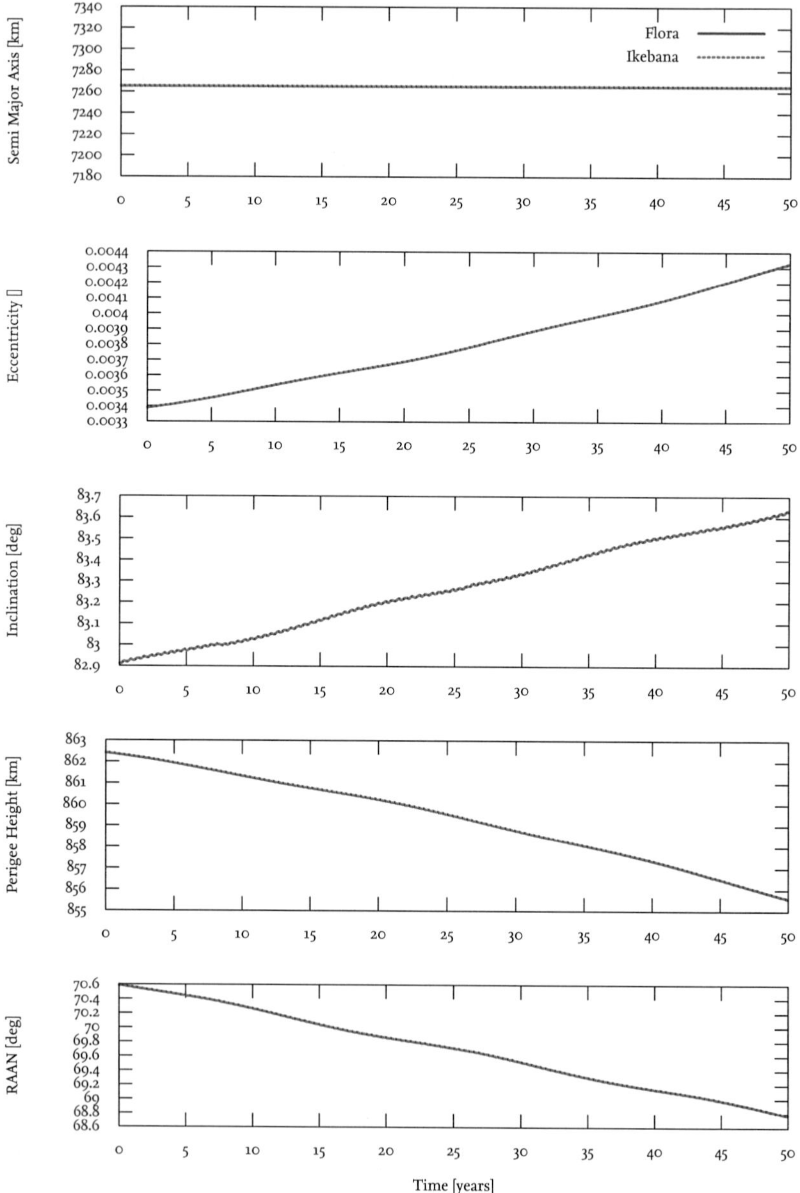

Figure 5.5.: Comparison of the third body perturbations module of Ikebana and FLORA on a LEO orbit. Object number: 13464

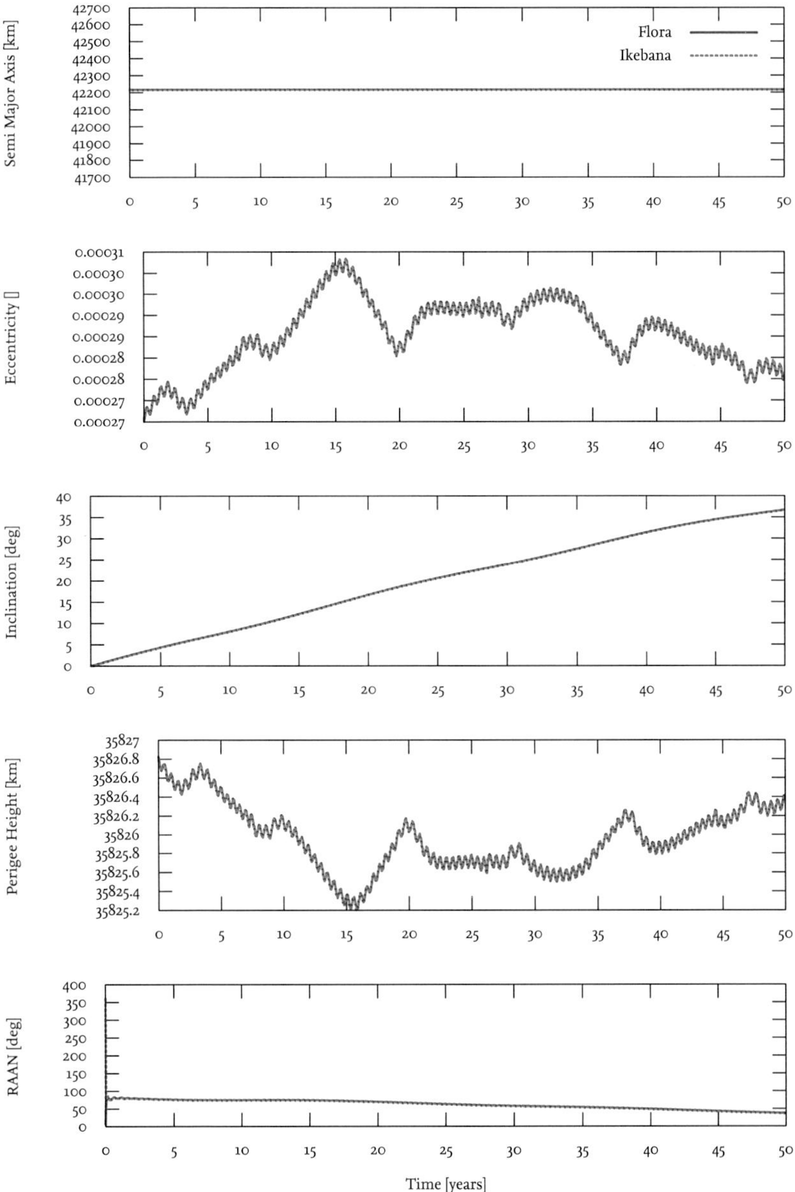

Figure 5.6.: Comparison of the third body perturbations module of Ikebana and FLORA on a GEO orbit. Object number: 28472

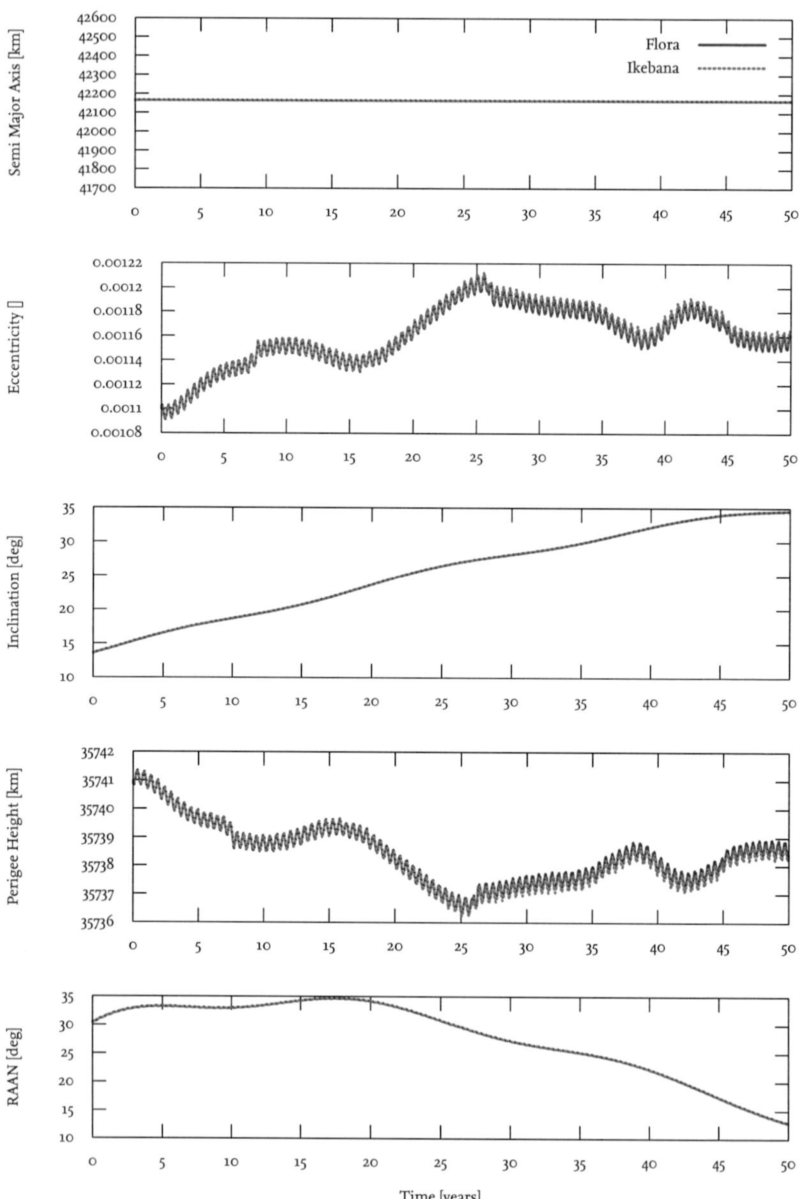

Figure 5.7.: Third body perturbations: Slight deviations in eccentricity caused by the code shown in listing 5.2. Object number: 22963

5.2.3.3. Solar Radiation Pressure

In the solar radiation pressure module, several effects can be observed. For many LEO orbits, Ikebana's result seems to oscillate around FLORA's value, sometimes with a slight increasing or decreasing tendency. This effect can be seen in figure 5.8. It's explanation lies in example given in section 5.2.1. Ikebana uses the second of the described methods to add changes to the original orbit, which results in these oscillations. The error caused by it is in the range of around ten metres.

In GTO orbits, the semi major axis of Ikebana sometimes deviates from FLORA in a more linear fashion; an example is shown in figure 5.9 where the values even propagate into different directions causing the error to increase over time. The maximum error sizes observed for this effect are in the range of 1-2 km over 50 years.

The other orbital elements often behave in a similar manner: In figure 5.8, this effect can be observed for inclination and RAAN, figure 5.9 also shows this for eccentricity. For the latter, the deviation occurs for eccentricities larger than 0.1; smaller eccentricities seem to be unaffected (figure 5.8). These deviations can be observed in all orbital regimes with error margins staying below 0.01 degrees for inclination, 0.1 degrees for RAAN and 0.0001 for eccentricity. A probable reason for this effect is the calculation of the ecliptic longitude of the Sun which heavily influences the solar radiation pressure acceleration and its direction. The value is calculated in double precision on the CPU but converted to single precision when it is uploaded to the GPU.

On very rare occasions, an anomaly can be observed when the eccentricity approaches zero. Some of the analytical equations that both FLORA and Ikebana are based on can cause a negative eccentricity to be generated. Both propagators work around this problem by artificially raising the eccentricity to a very small positive value. This introduces an error that FLORA and Ikebana handle differently. Usually the curves align after some steps or a small constant error is introduced. However, figure 5.10 shows that this problem can sometimes lead to unpredictable results that affect the other elements. Both FLORA and Ikebana show this behaviour, with Ikebana displaying it a little more frequently. While such occurrences are overall very rare the possibility of such occurrences must be taken into account.

In general the solar radiation pressure has very little effect on the catalogue objects. This is the reason why the deviations caused by Ikebana's lower floating point precision stay within an acceptable range. Since the force exerted on the object depends directly on its area-to-mass ratio, the influence of the perturbation and the severity of the error would multiply for high area-to-mass ratio (HAMR) objects. In addition, a step size of one day was used for the analysis which might be too large to accurately determine the shadow entry and exit points. For these reasons, an in-depth reevaluation of the underlying implementation in FLORA as well as Ikebana's use of single precision variables needs to be performed before a general conclusion can be derived.

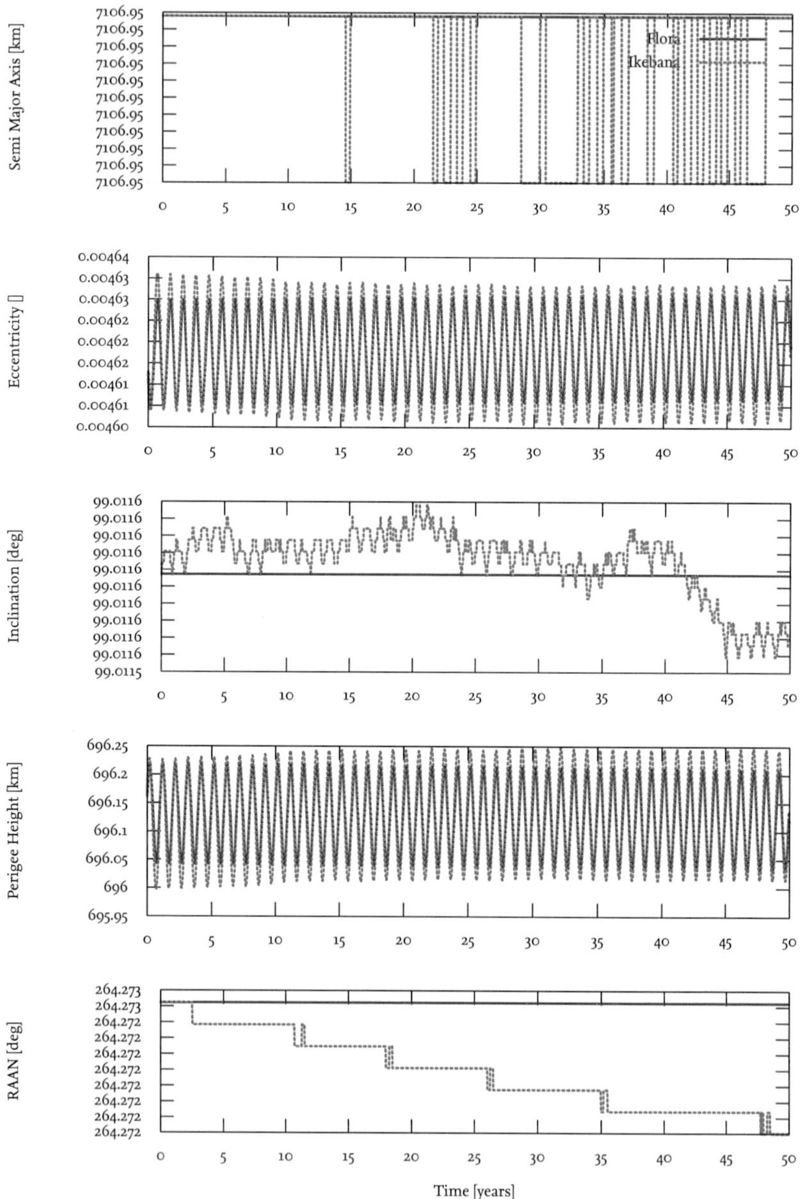

Figure 5.8.: Solar Radiation Pressure: This LEO orbit comparison shows several of the observed effects such as an oscillating semi major axis and small deviations in inclination and RAAN. Object number: 37452

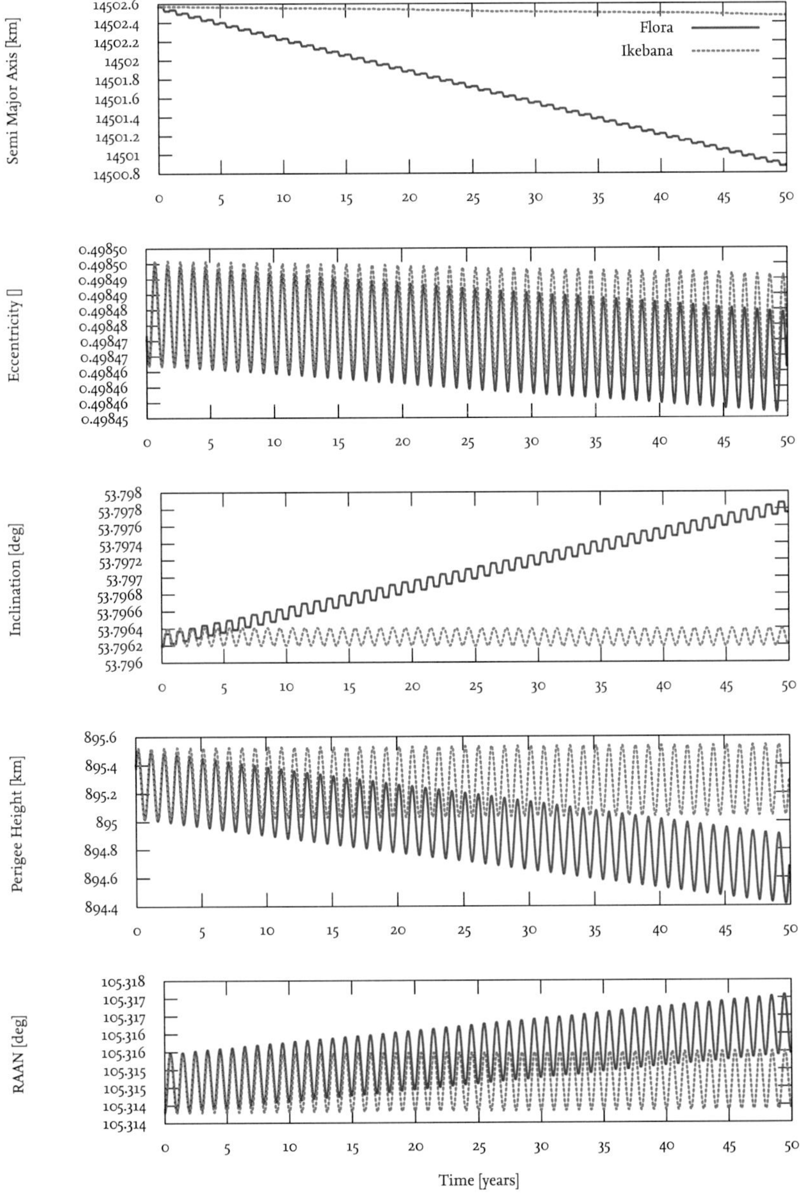

Figure 5.9.: Solar Radiation Pressure: Small deviations occur in all orbital elements. Object number: 34242

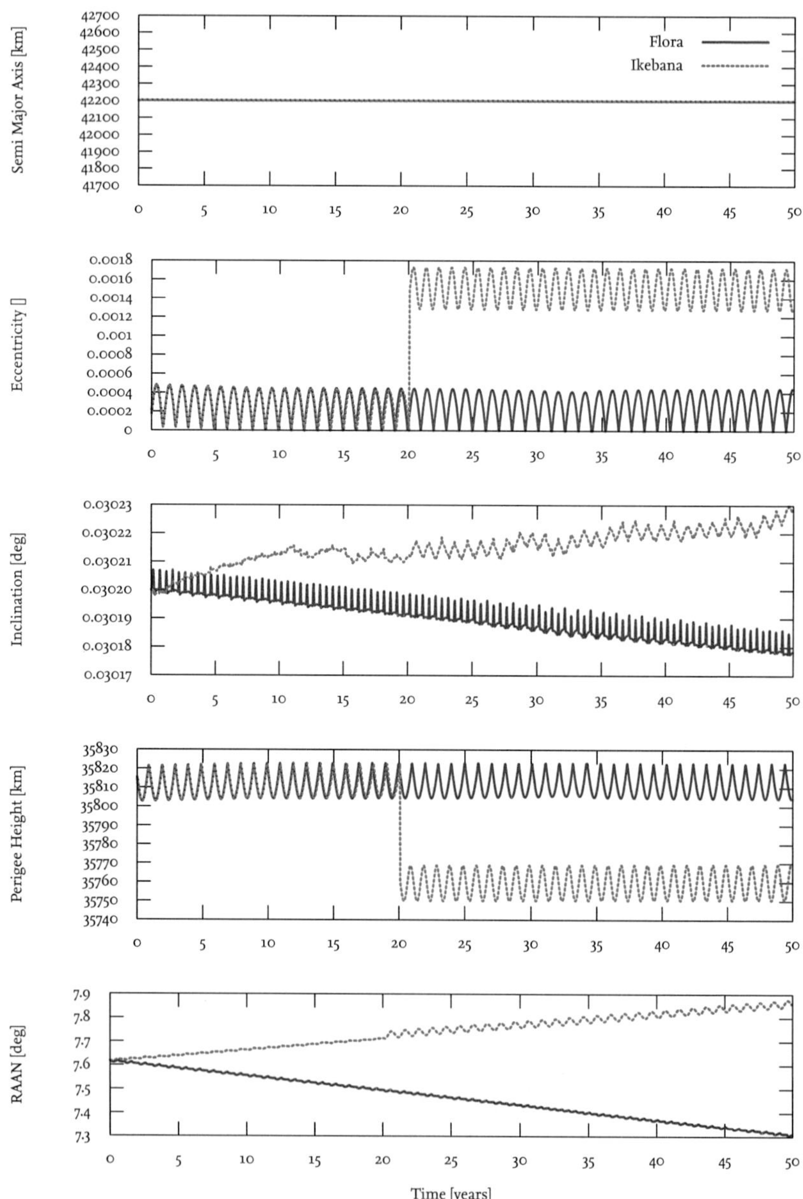

Figure 5.10.: Solar Radiation Pressure: Error handling for negative eccentricities can lead to unpredictable results. Object number: 38098

5.2.3.4. Atmospheric Model

The accuracy comparison of the atmospheric model shows good results. Obviously, objects with a perigee over 4000 km are not affected at all. For LEO regions, figure 5.11 shows typical deviations in eccentricity, inclination and RAAN. As in the solar radiation pressure module, these can be attributed to floating point inaccuracies and are within the same ranges for the examined population. Another common effect shown in that figure is the slightly stronger decrease of the semi major axis with an error of usually around 1-2 km. This deviation is caused by rounding errors in the interpolation of the atmospheric data values which result in slightly different scale height and density outputs. The error is assumed to be within the range of the uncertainties introduced by using a lookup table with a discretized height scale. It should be noted, however, that for low orbits such as the one shown in figure 5.12, this different behaviour leads to an earlier decay of the object and a slightly increased decay rate for the whole population (see table 5.3 in section 5.3.2).

Objects in GTO orbits usually show a similarly stronger decrease of the semi major axis (figure 5.13). GTO orbits with a very low perigee experience a strong atmospheric drag that causes the apogee to drop by several thousand kilometres after 50 years; in these cases the absolute deviation also increases into the 100 km range as seen in figure 5.14. This plot also shows that while Ikebana shows an overall higher decay rate, some objects also decay more slowly than in FLORA.

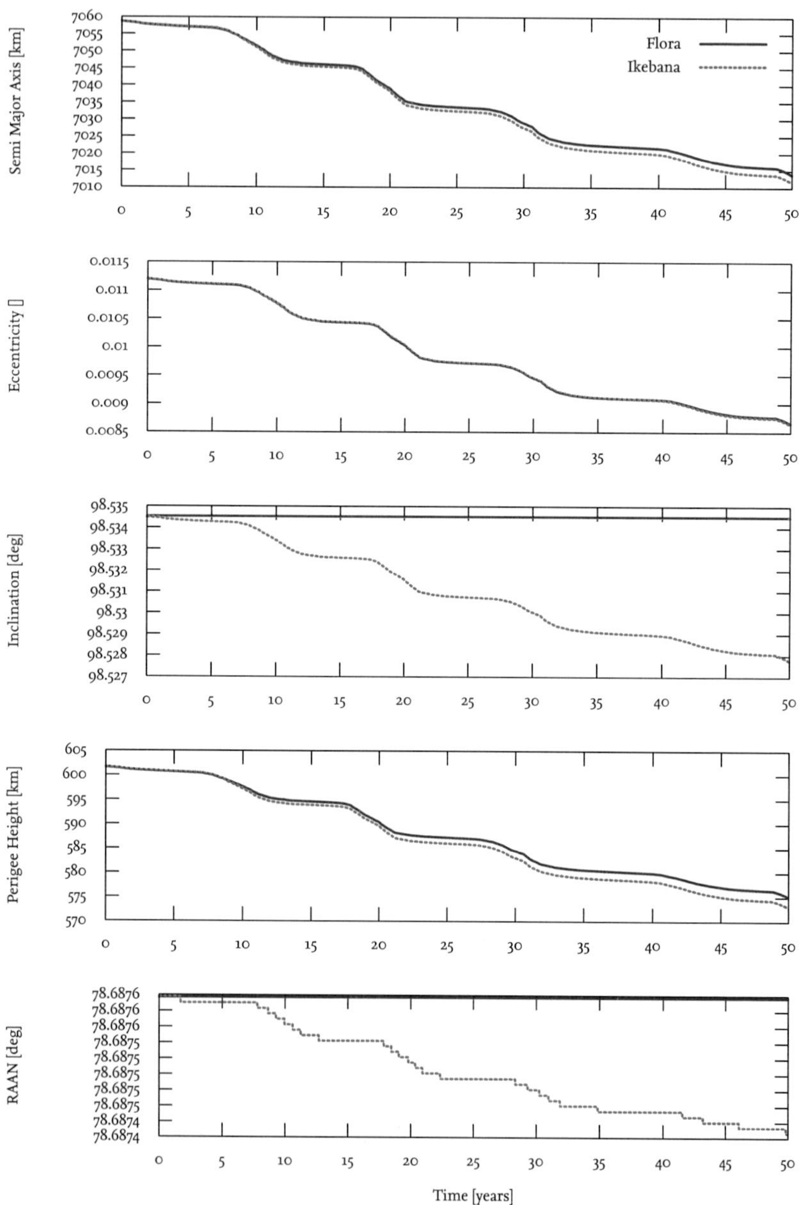

Figure 5.11.: Atmosphere: Slight deviations in eccentricity, inclination, and RAAN. Object number: 28075

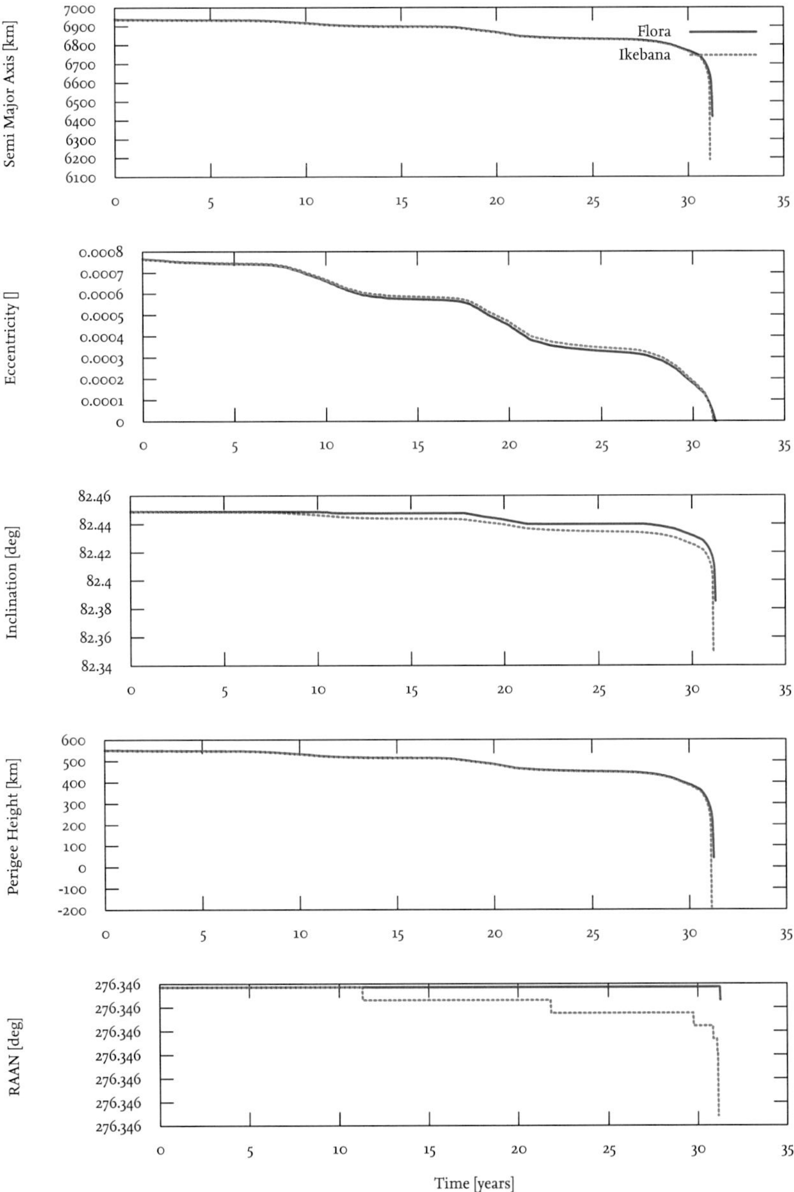

Figure 5.12.: Atmosphere: The semi major axis declines slightly faster in Ikebana. Object number: 33504

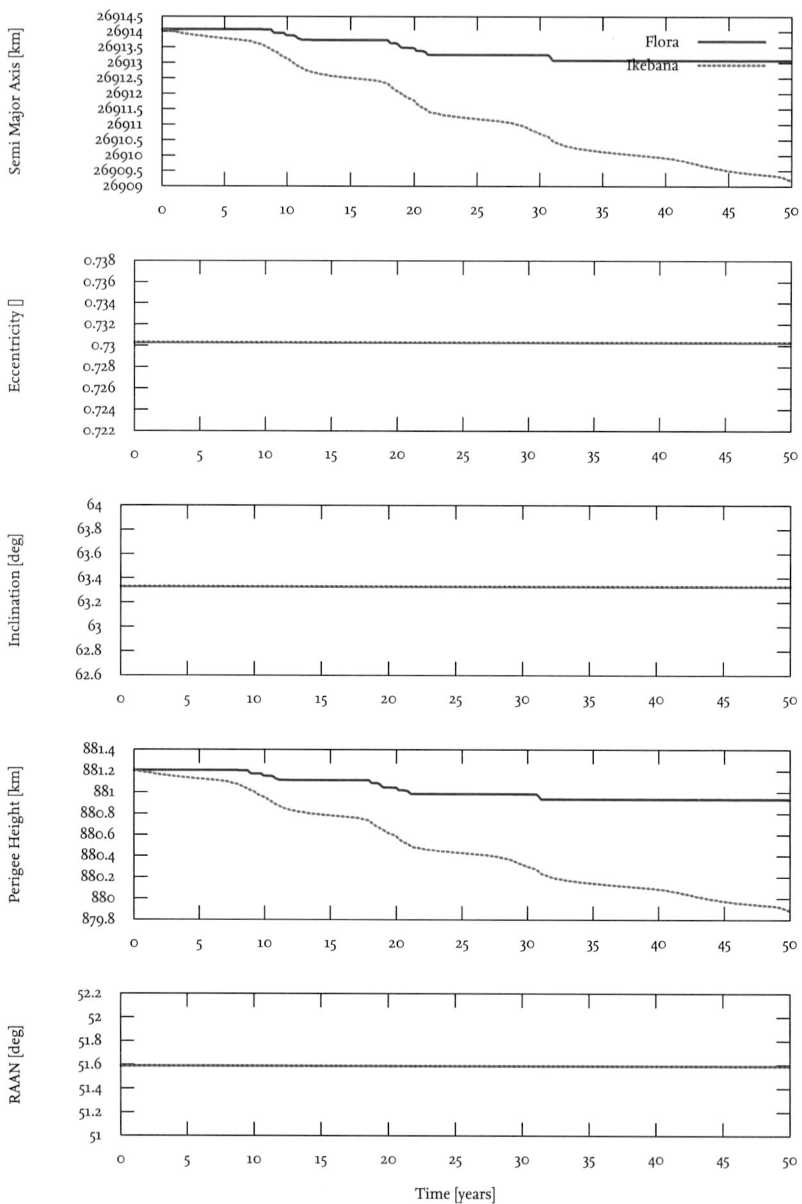

Figure 5.13.: Atmosphere: Decline of the semi major axis in high-eccentricity GTO orbits Object number: 25850

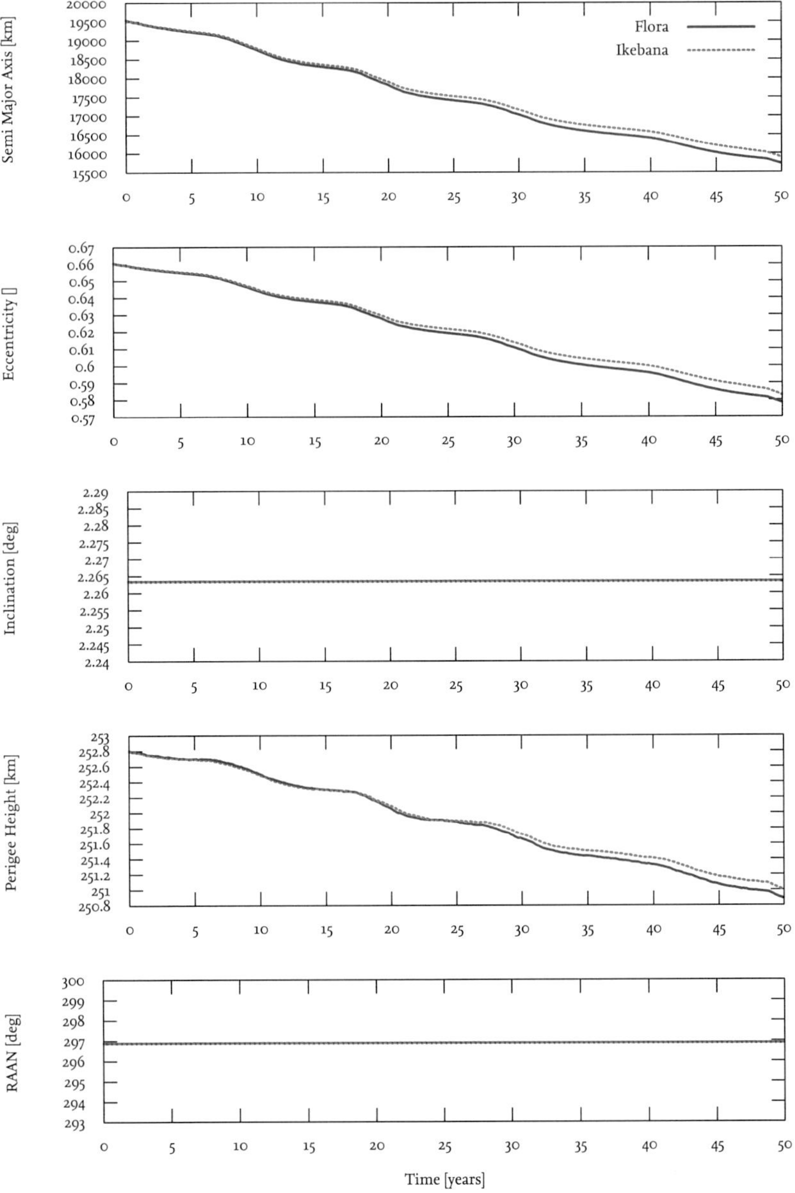

Figure 5.14.: Atmosphere: Decline of the semi major axis in high-eccentricity GTO orbits Object number: 36833

5.2.4. Total Accuracy Results

Adding the perturbed orbits of all four models provides good overall results. Figure 5.19 shows a typical LEO orbit with the slightly faster decline of the semi major axis caused by the atmosphere and only very minor changes to the other orbital elements. The faster decay rate shown for the atmospheric model is reflected in the overall results as well (figure 5.20). Like in the individual model results, the relative errors of semi major axis, eccentricity and inclination are generally acceptable for statistical analysis of the space debris population. Figures 5.15 and 5.18 show an overview of the errors in the orbital elements Ikebana produced compared to FLORA; table 5.2 lists the classifications used for the histogram plot. The definition of the border between "acceptable" and "large" is based on experience with long-term evaluation of large populations; it is placed at the point where the deviations are deemed to cause significant statistical discrepancies if a high number of objects is affected.

Element	Very small	Small	Acceptable	Large
Semi Major Axis	$< 2km$	$< 10km$	$< 100km$	$\geq 100km$
Eccentricity	< 0.0001	< 0.001	< 0.01	≥ 0.01
Inclination	$< 0.001°$	$< 0.01°$	$< 0.1°$	$\geq 0.1°$
Perigee Height	$< 2km$	$< 10km$	$< 100km$	$\geq 100km$
RAAN	$< 0.1°$	$< 1°$	$< 10°$	$\geq 10°$

Table 5.2.: Classification of the results.

Over a propagation time frame of 50 years, the deviation of the semi major axis stays below 2 km for 90 per cent of the objects and under 10 km for 96 per cent. For the eccentricity, the error stays below 0.0001 for 88 per cent of the objects and under 0.001 for 98 per cent. Inclination deviations are lower than 0.001 degrees for only 36 per cent of the objects but 98 per cent stay under 0.1 degrees. The relatively fast-changing right ascension of the ascending node shows deviations of less than ten degrees for 84 per cent of the objects. While this might cause a relatively large positional error for some objects it is acceptable for statistical analysis where the exact position of the object is not important.

LEO objects with an error of more than 10 km in semi major axis are often those for which decay is imminent; since Ikebana generally shows earlier decay than FLORA, the deviations increase during this state (figure 5.21). Orbits with strong deviations in semi major axis and eccentricity are highly eccentric orbits with a low perigee and a high apogee such as GTO and Molniya orbits that experience a strong atmospheric drag in the perigee region (figure 5.22). In almost all cases, the atmospheric model is responsible for the largest deviations. The error plot with the atmospherical model disabled is given in appendix B and shows reduced error rates in semi major axis and eccentricity by around two orders of magnitude (figure B.5). For the above examples, results with the atmospherical model disabled are shown in figure 5.23. Thus, GEO orbits with no atmospherical disturbances are generally free of errors; a typical example is shown in figure 5.24. The long-periodic perturbations visible in inclination and eccentricity are reproduced accurately. The errors introduced in the solar radiation pressure module which propagated some of the orbital elements into opposing directions are small enough to be cancelled out when changes from other modules are added to the delta orbit. This can be seen in figure 5.25 when compared to figure 5.9. In cases where the solar radiation pressure is the only perturbation force that is taken into account, this effect should be further investigated. The above figure also shows the slight deviation of the semi major axis that is present in many GTO orbits.

As stated above, this data gives only a relative comparison between two analytical propagators and provides no evidence for the absolute quality of either FLORA or Ikebana. FLORA has been validated in the process of a long-term population analysis project ([Möckel et al., 2013]) against the numerical propagator ZUNIEM and was found to provide satisfactory results. However, the report also mentions that the solar radiation pressure model built into FLORA shows the largest relative errors and works on the LEO population only because the influence of the SRP perturbation on catalogue objects is very small. Since Ikebana amplifies this effect the model's implementation should be reevaluated in both propagators. Otherwise, the deviations that Ikebana introduces are within the range of the uncertainties that are intrinsic to analytical propagation. According to [Tapping and Charrois, 1993], errors in determining the F10.7 values are in the range of around 2%. [Bruinsma et al., 2012] compared the density output of the NRLMSISE-00 model to actual measurement data and found deviations between 3% and 10% for different sets of data. Figures 5.16 and 5.17 show the error rates introduced into FLORA when F10.7 and density values are overshot by two and three per cent, respectively. Comparing this to figure 5.15 shows that the error introduced by precision loss in Ikebana is in the same order. The slightly higher deviations in RAAN and inclination can be explained by the fact that the first figure includes differences from all perturbation models whereas the parameter variations in FLORA only affect the atmospheric model.

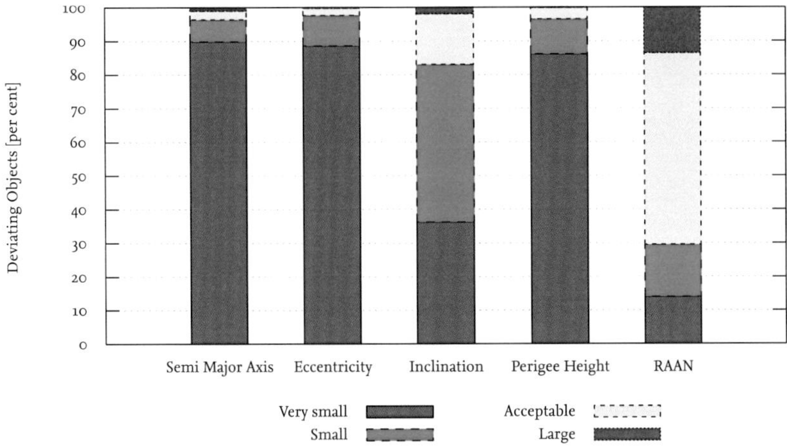

Figure 5.15.: Total Results: Histogram showing the error rates between FLORA and Ikebana.

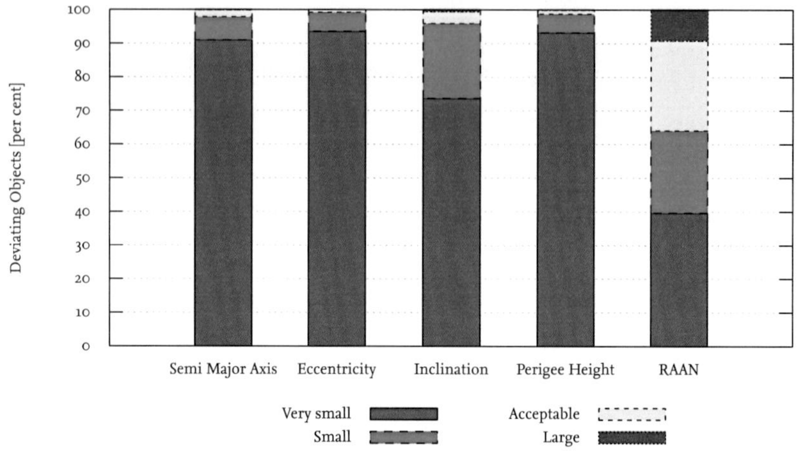

Figure 5.16.: Total Results: Histogram showing the error introduced into FLORA's output when the F10.7 values for solar activity are overestimated by 2%.

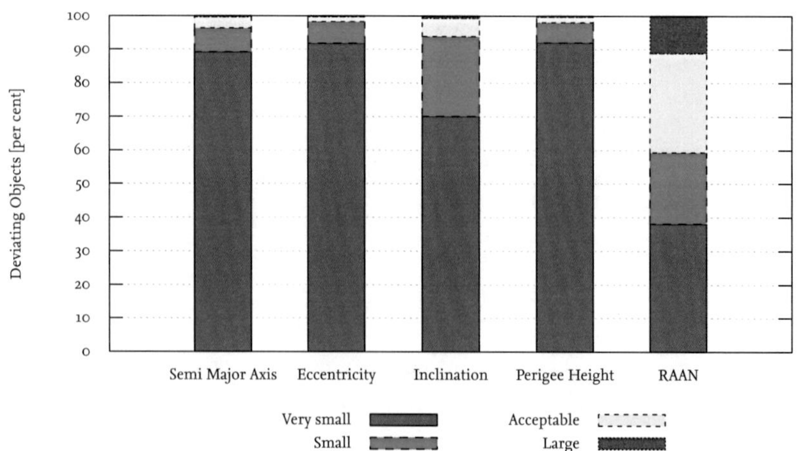

Figure 5.17.: Total Results: Histogram showing the error introduced into FLORA's output when the atmospheric density values are overestimated by 3%.

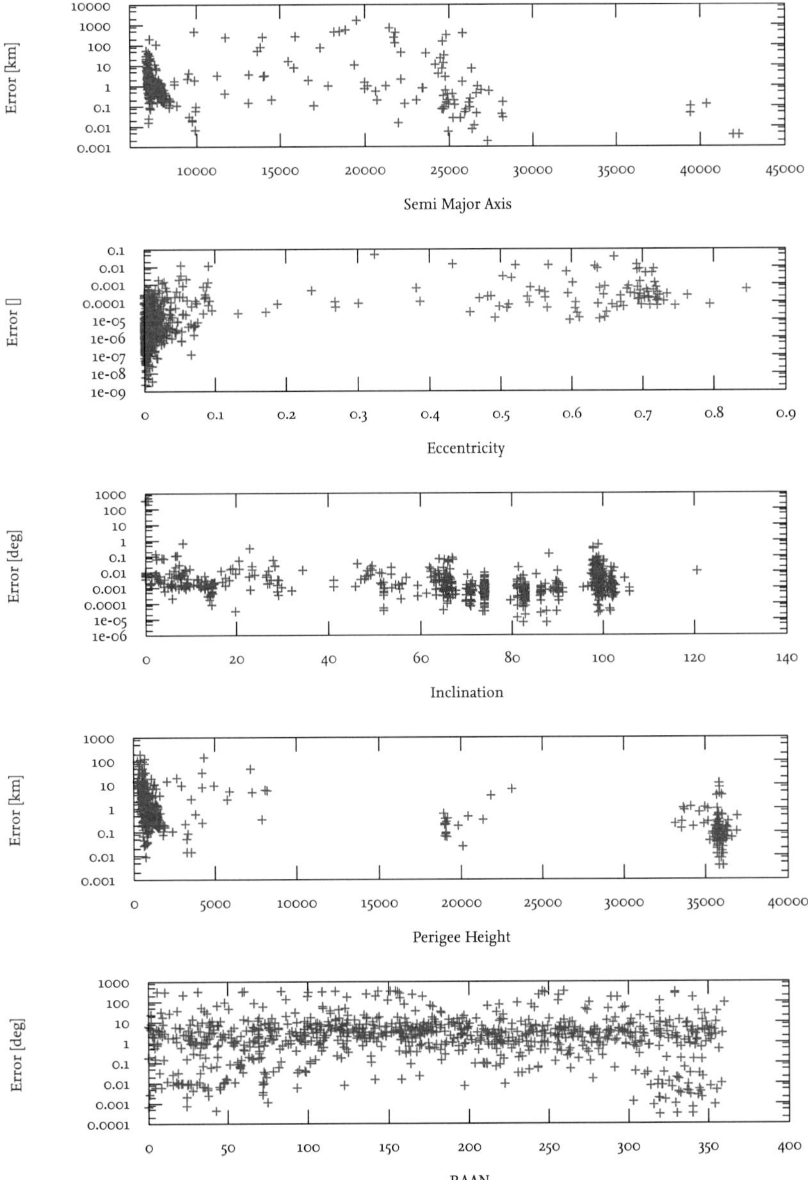

Figure 5.18.: Total Results: Error rates of 1000 random objects.

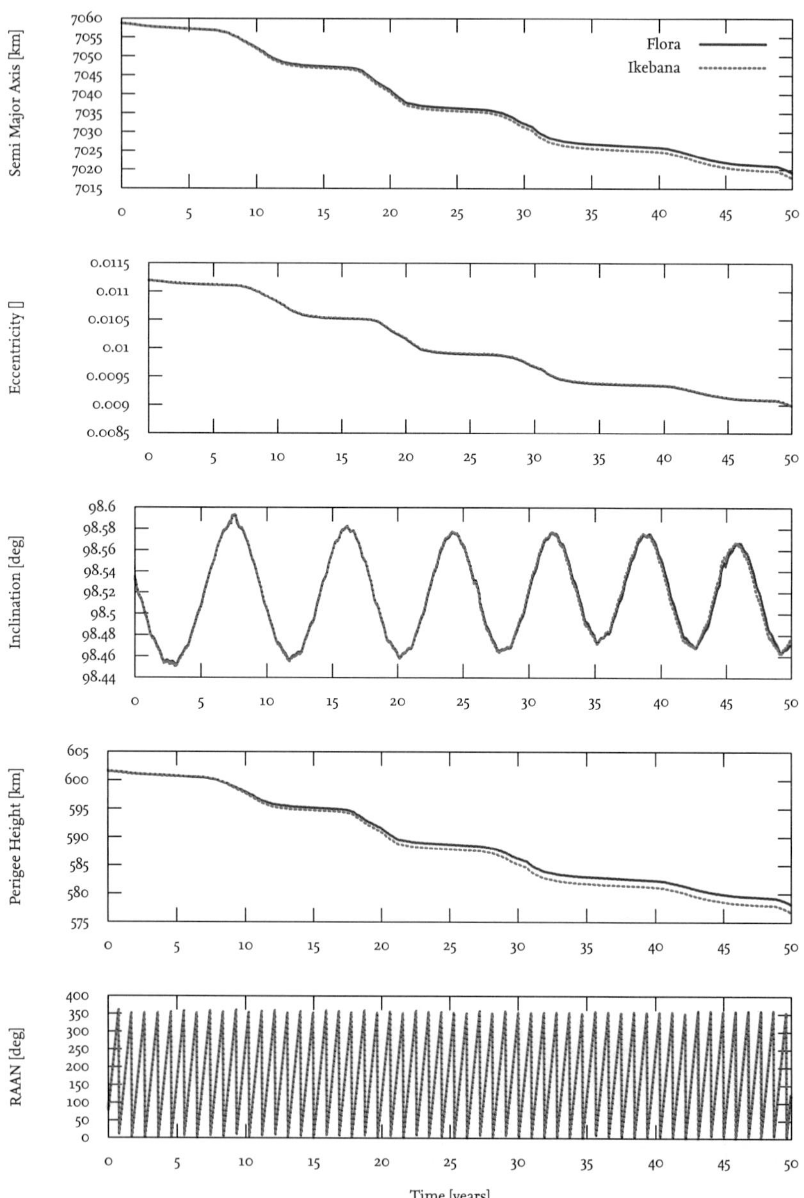

Figure 5.19.: Total Results: Typical LEO object with slightly faster decline of the semi major axis and no major changes in the other elements. Object number: 28075

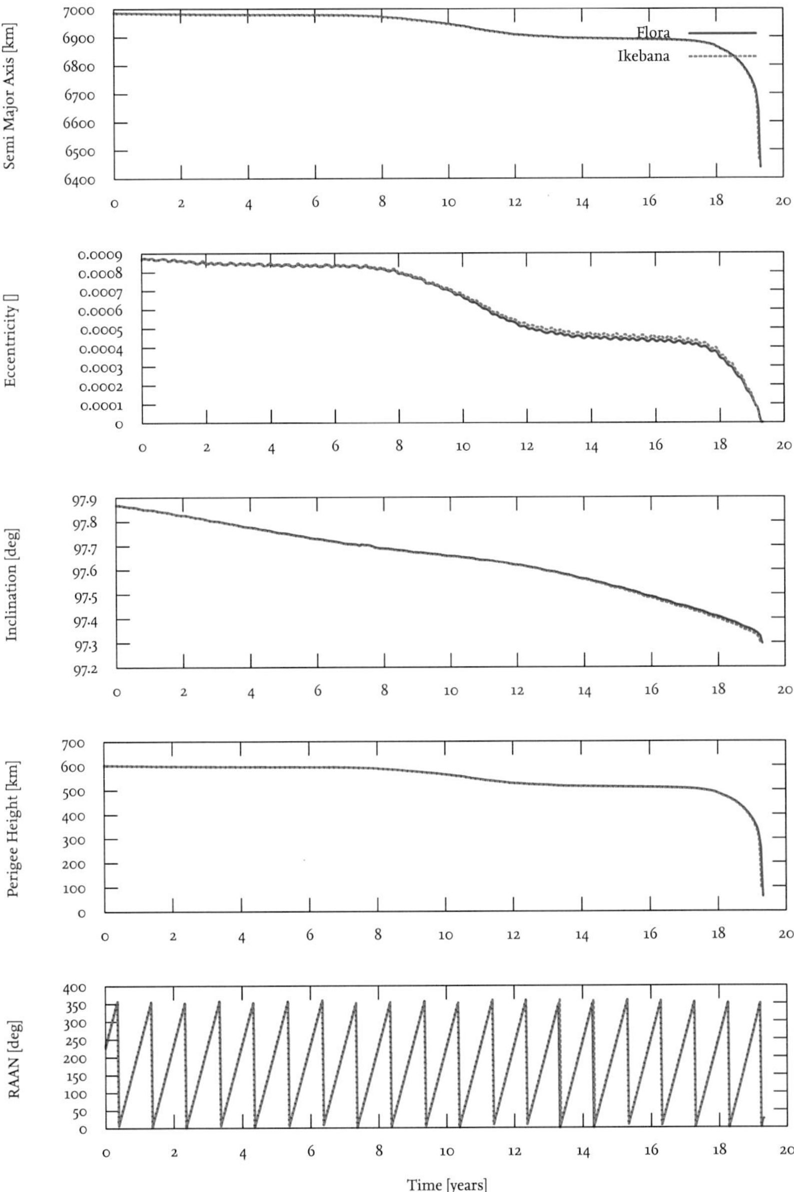

Figure 5.20.: Total Results: The faster decline of the semi major axis in the atmospherical model causes an overall slightly increased decay rate. Object number: 39769

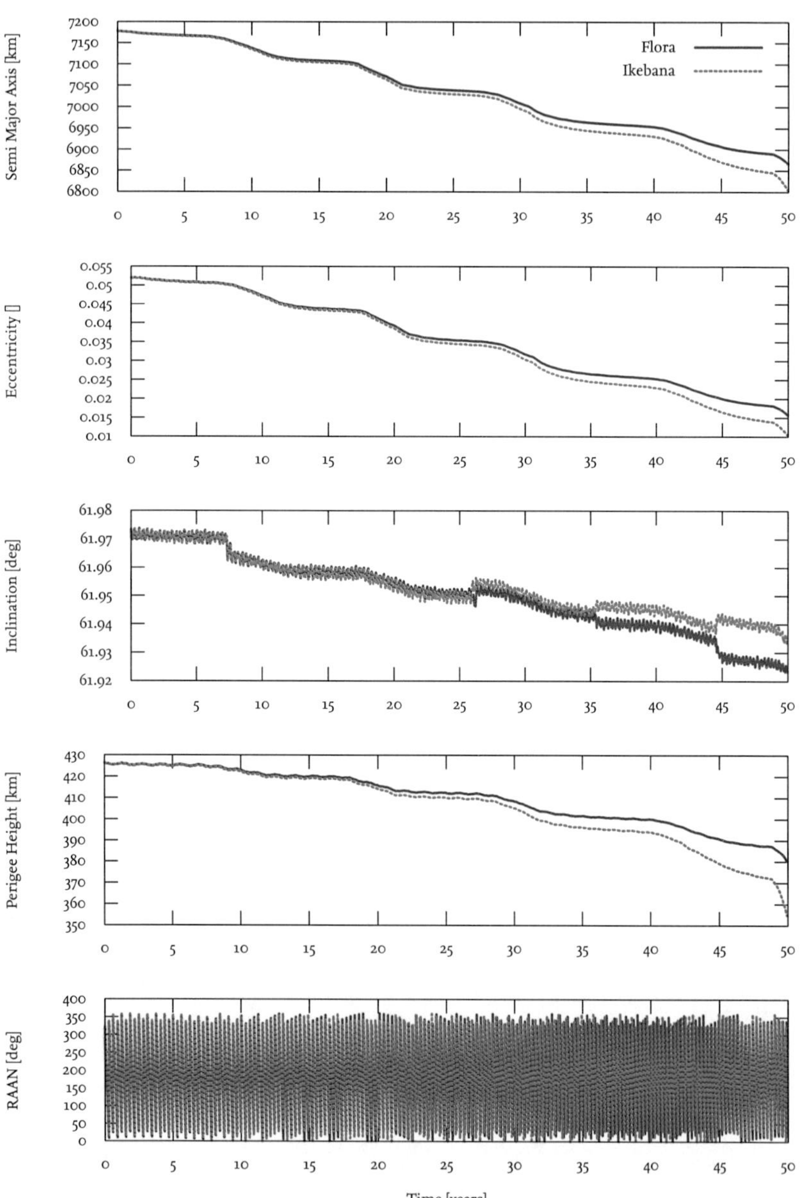

Figure 5.21.: Total Results: LEO objects with imminent decay show relatively large deviations of semi major axis and eccentricity at the final data point. Object number: 3757

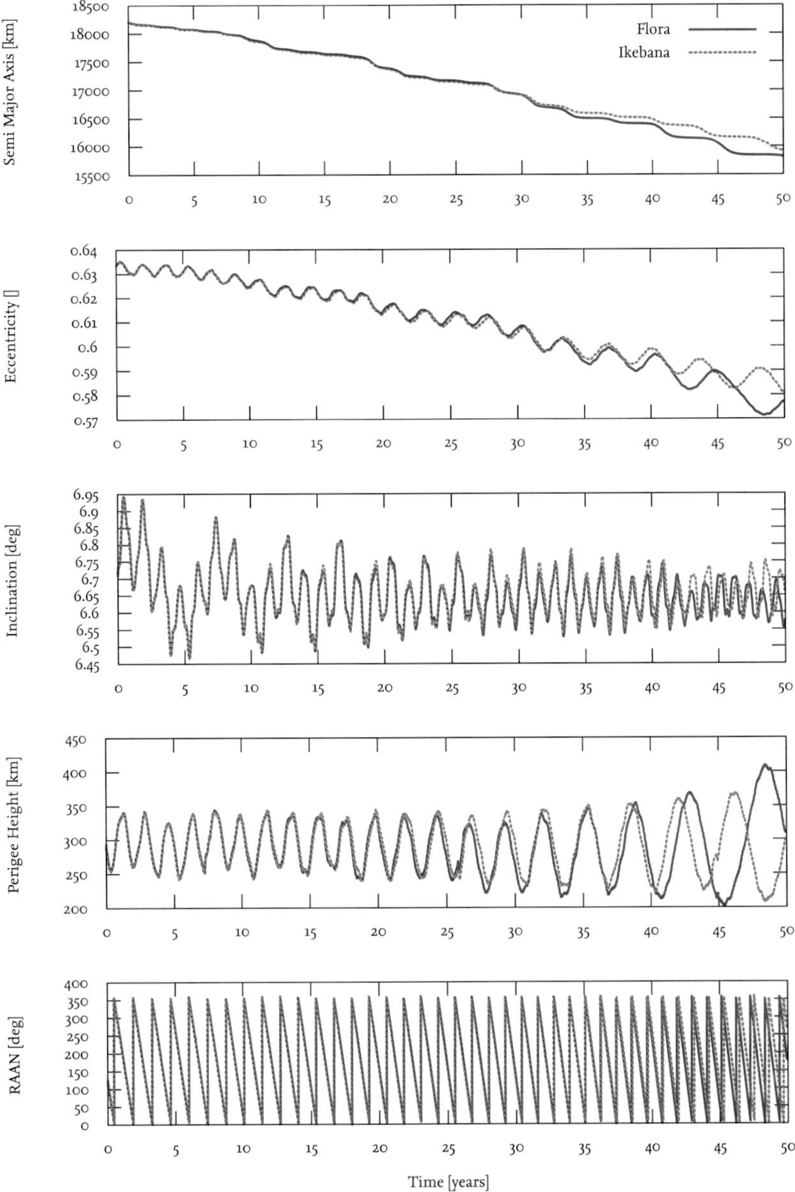

Figure 5.22.: Total Results: Some highly eccentric orbits with low perigees show relatively large deviations of semi major axis and eccentricity. Object number: 22670

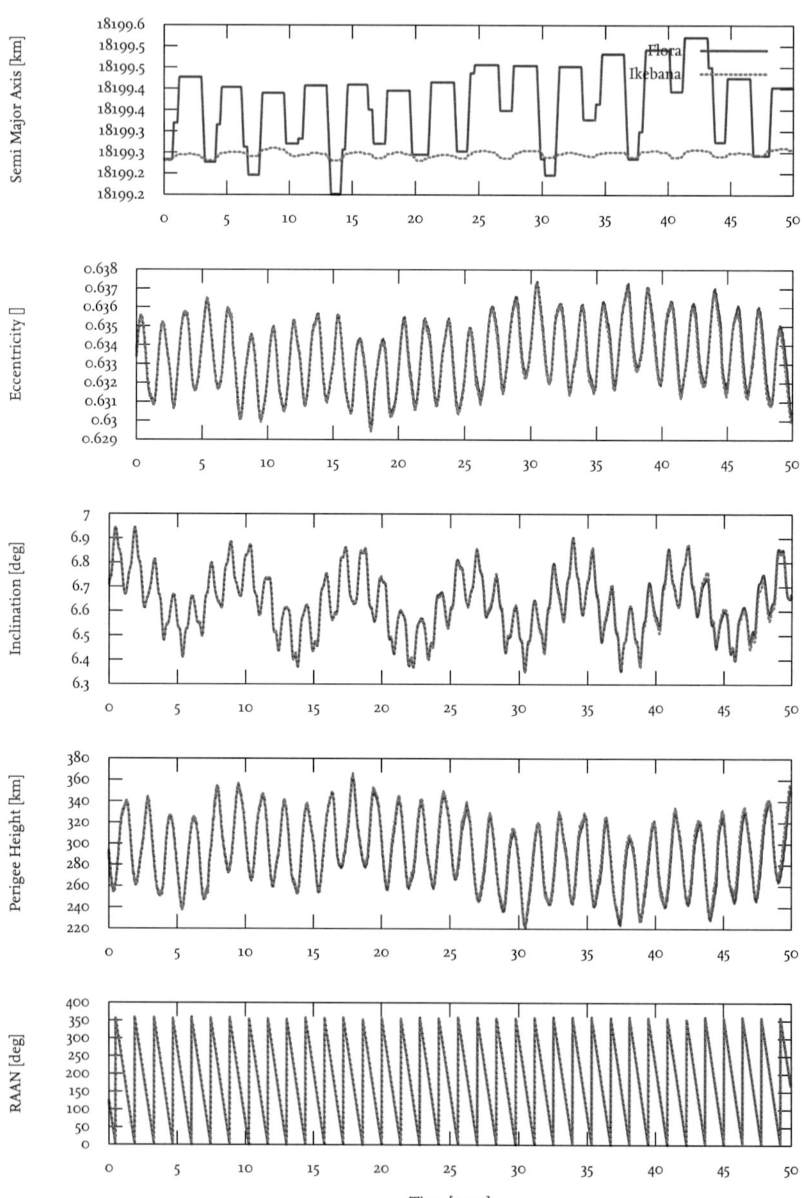

Figure 5.23.: Total Results: Without the atmospherical model, the inaccuracies visible in figure 5.22 have largely vanished. Object number: 22670

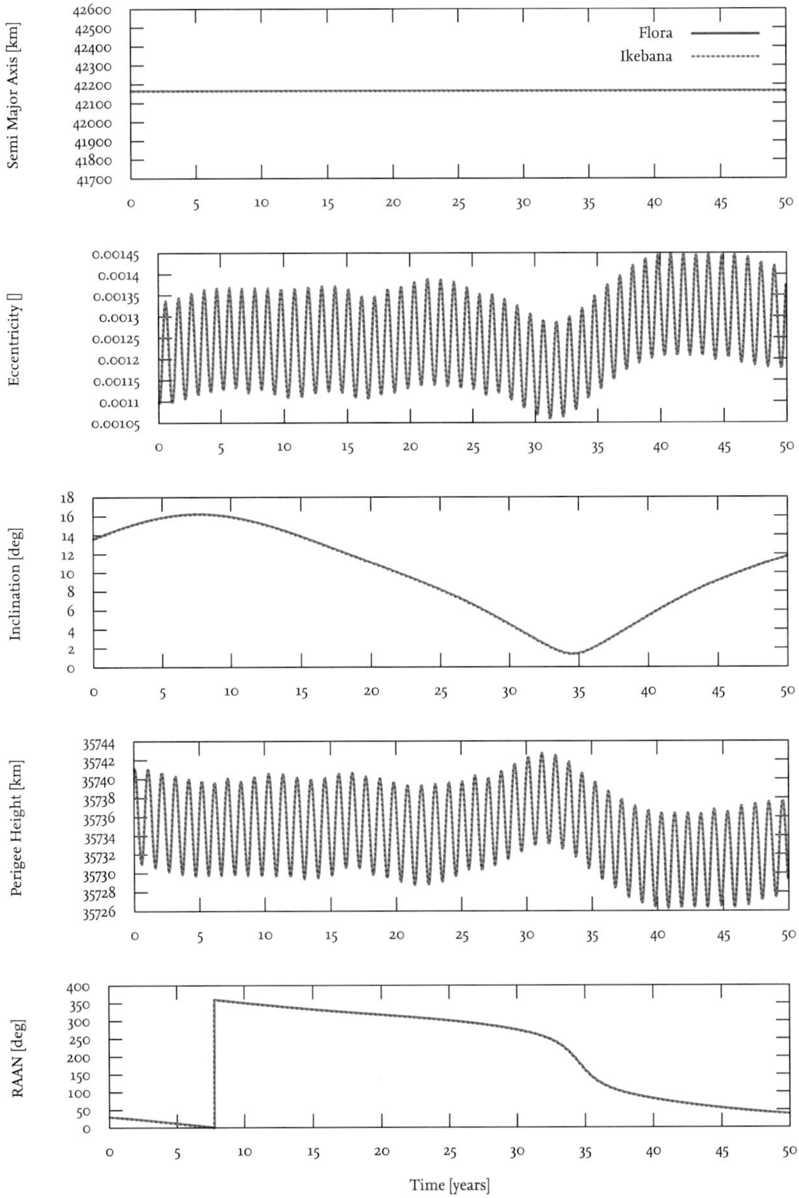

Figure 5.24.: Total Results: Propagation of GEO orbits is generally free of deviations. Object number: 22963

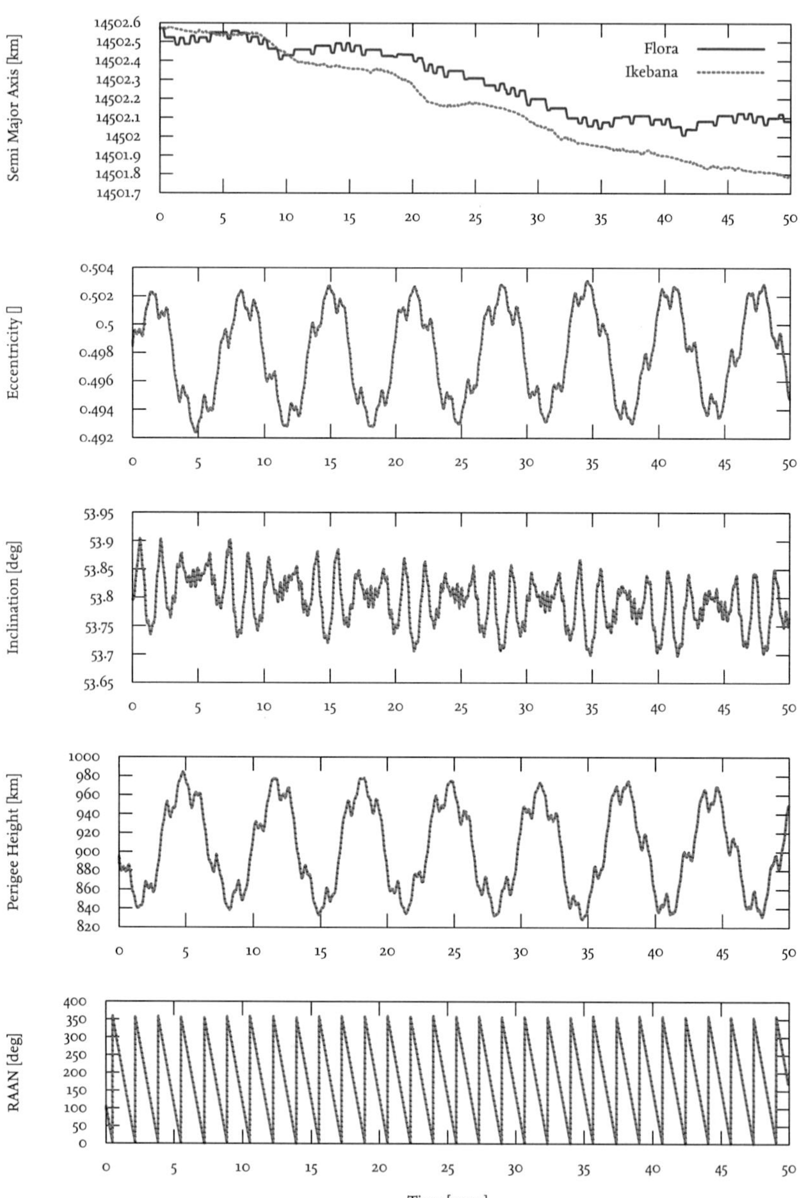

Figure 5.25.: Total Results: The deviations introduced by the solar radiation pressure module (compare figure 5.9) are small enough to get cancelled out. Object number: 34242

5.2.5. TLE Data Comparison

To give an idea of FLORA's and Ikebana's absolute accuracy, both propagators were tested against the publicly available TLE data set of *Vanguard-1* (figure 5.26). This satellite was launched in 1958 and has not yet decayed, making it the oldest man-made satellite still in orbit and one of the few objects for which positional data is available for over 50 years. TLE data was aggregated for dates between August 11, 1964 and December 28, 2014; FLORA and Ikebana were configured to propagate the initial set of orbital elements for 50 years with a time step of 12 hours. The satellite has a diameter of 0.165m and a mass of 1.47kg which amounts to an area-to-mass ratio of 0.0145 square metres per kilogram. For the drag coefficient, the default value of 2.2 was assumed. The resulting orbital data is shown in figure 5.28. After 50 years, both FLORA and Ikebana show deviations in semi major axis of around 20 km and in eccentricity of around 0.002. While the decay of the original orbit is slightly faster than what the propagators predict, the overall progress is reproduced accurately. For the inclination, FLORA and Ikebana show a smaller amplitude in the periodic disturbances but stay well within the average range. The perigee height shows minor deviations but again, the overall progess is depicted accurately. In all cases, the data shows that the floating point deviations between FLORA and Ikebana are negligible compared to the uncertainties that are inherent to the propagation algorithm itself as well as the input values. To demonstrate the latter, figure 5.27 shows an additional propagation with a drag coefficient of 2.4 (see [Bowman, 2002]). The deviations caused by this change slightly improve the absolute results and are more significant than the difference between FLORA and Ikebana.

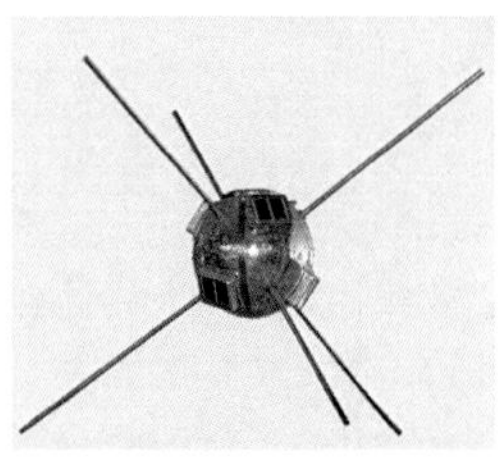

Figure 5.26.: *Vanguard-1*.

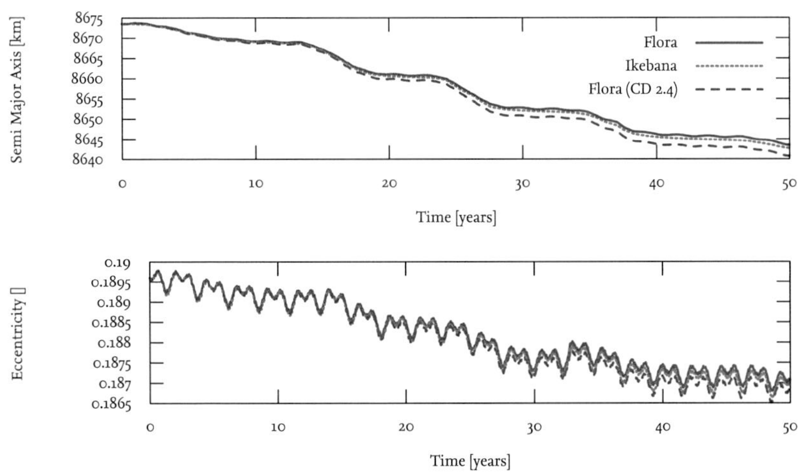

Figure 5.27.: Vanguard-1 propagated with FLORA and Ikebana. The dashed line shows the changes caused by setting the drag coefficient to 2.4 which is realistic according to [Bowman, 2002]).

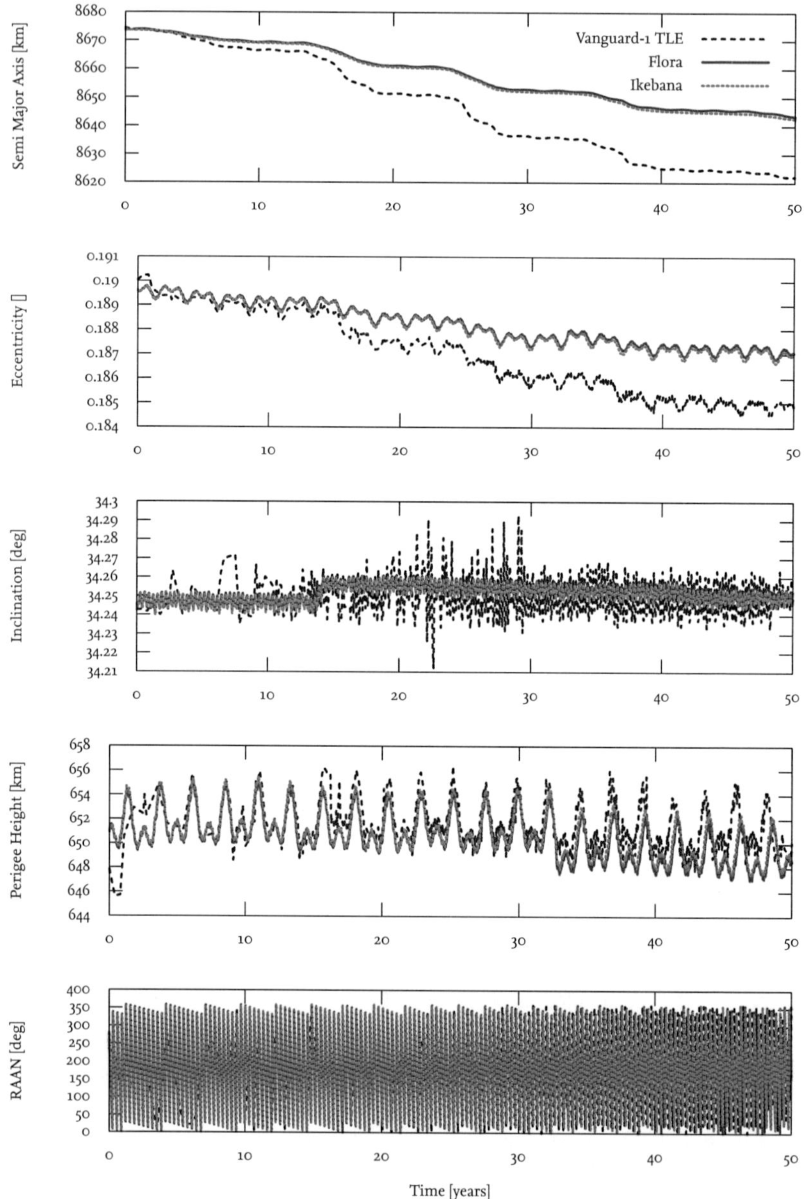

Figure 5.28.: Vanguard-1 TLE data plotted against FLORA and Ikebana results.

5.3. Speed

5.3.1. Runtime Evaluation

The process of propagating a given set of objects multiple times over a given time frame is called a propagation scenario. The time required for computing a propagation scenario with a sequential, analytical propagator depends on three main factors:

- N_{obj} - the number of objects to be propagated
- N_{st} - the number of time steps to be propagated
- t_{st} - the average computation time for one object time step

If any overhead such as loading data into memory or writing output to a hard drive is neglected the overall run time for a scenario is the product of these three values:

$$t_{total} = N_{obj} \cdot N_{st} \cdot t_{st} \tag{5.1}$$

The number of time steps is usually broken down into a propagation time frame T_P which is divided into equally-sized segments, the propagation step size ΔT:

$$N_{st} = \frac{T_P}{\Delta T} \tag{5.2}$$

The product of N_{obj} and N_{st} gives the number of propagation operations, N_P. This is exactly the number of times the propagation algorithm is called for the given scenario:

$$N_P = N_{obj} \cdot N_{st} \tag{5.3}$$

In the case of ideal parallel propagation, t_{st} is divided by the number of objects that can be processed at once resulting in a speedup factor equal to the number of parallel threads. In reality, complete parallelization is not possible or useful. In Ikebana, for example, this applies to everything that happens in the *setTimeStep* functions. The equation describing the maximum achievable speedup as a function of the number of processors, N_{proc}, is known as *Amdahl's Law* ([Amdahl, 1967]):

$$S(N_{proc}) = \frac{1}{(1 - P_{par}) + \frac{P_{par}}{N_{proc}}} \tag{5.4}$$

where P_{par} is the percentage of the algorithm that can be parallelized. It increases with larger population sizes and smaller numbers of time steps.

The values for N_{obj}, T_P and ΔT are usually given through the definition of the propagation scenario. During a long-term analysis, for example, a population of a given size is used as a starting point. T_P is the time frame for which the population should be examined, e.g. 200 years from a given date for which the initial population is valid. ΔT is used to configure the desired trade-off between accuracy and run-time requirements and can be anywhere between a millisecond and several years. As a rule of thumb, the smaller the value for ΔT, the more accurate the results will be but the computation time will increase linearly according to eq. 5.1. In the visualization use case, N_{obj} is the size of the population that is being displayed and T_P is the length of the animation, usually in the range of several minutes. ΔT is adjusted dynamically at run-time; it is linked to the configured animation speed and the program's frame rate to ensure a fluid animation, usually ranging between several milliseconds to several minutes. The value for t_{st} depends on a lot of factors such as the computer's processing power and workload, the complexity of the perturbation models and the properties of the population objects that state, for example, if and how long an object will travel through the Earth's atmosphere. Therefore, t_{st} is determined by measuring t_{total} of a reference population and taking the average over all objects and time steps. Keeping t_{total} as small as possible is the goal of optimizing the run time of a propagator.

Example: For a long-term analysis study ([Möckel et al., 2015]), several scenarios have been executed for a time frame of 200 years between 2013 and 2213 with a population of 49,000 objects[1]. With the step size ΔT set to three days, the overall time required for computing this scenario on a single core of a typical desktop PC was measured to be a little under two hours.

$$t_{total} = N_{obj} \cdot N_{st} \cdot t_{st} = 49,000 \cdot \frac{200 \cdot 365.25d}{3d} \cdot t_{st} = 7,159s \approx 1.99h \tag{5.5}$$

This gives an average t_{st} of about six microseconds - circa 21,600 CPU cycles on a 3.6GHz processor:

$$t_{st} = \frac{t_{total}}{N_{obj} \cdot N_{st}} = \frac{7,159s}{49,000 \cdot \frac{200 \cdot 365.25d}{3d}} \approx 6 \cdot 10^{-6}s \tag{5.6}$$

5.3.2. Benchmarking

As an easily comparable and convenient index for propagation algorithms the ratio between N_P and t_{total} can be used, multiplied with a factor of 10^{-6} for readability. Thus, the *Propagator Benchmark Index*, $\check{M}$, is defined as

$$\check{M} = \frac{1}{t_{st} \cdot 10^6} = \frac{N_P}{t_{total} \cdot 10^6} \tag{5.7}$$

For the above example, the index can be calculated as

$$\check{M} = \frac{N_P}{t_{total} \cdot 10^6} = \frac{49,000 \cdot \frac{200 \cdot 365.25d}{3d}}{7,159s \cdot 10^6} = 0.167 \frac{MP}{s} \tag{5.8}$$

or *megapropagations per second*. From the two possible candidates, the other being t_{total}, this value was chosen as a benchmark because of its independence from object count and number of time steps as well as its clear definition: While the term "run time" can refer to anything from total propagator run time, total host run time, propagation with or without data transfer, etc., "(mega)propagations per second" defines exactly what is being measured. With $\check{M}$ being the inverse of t_{st} the benchmark depends on the same uncertainties as this variable. It is also important to note that t_{st} is averaged over the whole population. In reality, an object that is outside the atmosphere will have a much shorter propagation time than one for which the atmospherical disturbances have to be calculated. The propagation time for an object that has already been marked as decayed will, ideally, be no longer than it takes the processor so evaluate a single *if*-statement. Propagating a population consisting of only GEO or fast-decaying objects will therefore always be faster than a LEO population of the same size.

[1] In the actual scenario, the object number increased over time due to launches and collisions. For this example, a fixed object number is assumed.

To compare the run time of two propagators it is necessary to minimize these uncertainties by ensuring the following conditions:

The population and configuration given to both propagators must be identical. For the reasons stated in the above example, this is the most important condition. The population should also be representative of the propagator's specific use case.

Both propagators must achieve the same level of accuracy. Again, the question of how much accuracy is sufficient depends on the propagator's use case. If two algorithms are sufficiently accurate for a specific task a benchmark test can be performed to identify and select the faster one.

Both propagators must lead to a similar decay rate. Objects that have been marked as decayed will be excluded from the propagation process. Therefore, a propagator that shows a higher decay rate will likely be faster because the number of full propagation operations is smaller. Table 5.3 shows the mean decay rates of FLORA and Ikebana over the time periods that were chosen for the benchmark. Since Ikebana was aimed at recreating the results of FLORA as accurately as possible, the decay rates of the two propagators would ideally be identical. In fact, Ikebana shows a slight tendency towards earlier decays that can be attributed to slight deviations in the atmospherical model as described previously. The difference is however not large enough to have a significant influence on the run times. In an earlier version of the application, a programming error caused the decay rate of Ikebana to be slightly less than FLORA's; the effect on the benchmark results was negligible, and the error did not otherwise affect the propagator's run time.

$\frac{N_P}{t_{total}}$	**FLORA**	**Ikebana**
1 year	0.94%	0.99%
5 years	2.09%	2.19%
10 years	4.02%	4.23%
20 years	7.45%	7.76%
50 years	12.06%	12.47%

Table 5.3.: Mean decay rates of FLORA and Ikebana over different propagation times.

Both propagators must be executed on the same platform. Usually this condition is to ensure that no platform-specific bottlenecks falsify the results. However, when comparing a GPU adaptation of a CPU algorithm, this is inherently impossible. For this reason benchmarks are often given using another layer of abstraction such as "performance per dollar" or "performance per watt", although the usefulness of the latter has been questioned ([Akenine-Möller and Johnsson, 2012]). Since this work focuses mainly on practical use of the GPU propagator any further abstraction is omitted. Instead, performance tests were conducted on various off-the-shelf graphics cards that are readily available in an office environment or can easily be installed into a standard PC. The hardware choices are outlined in the next chapter with power consumption and approximate price added for reference.

5.3.3. Performance Evaluation Setup

To analyze the performance of Ikebana versus FLORA, the reference population detailed in section 5.1 was used as a basis. From the randomized population, subsets of 1000, 10,000,

50,000, 100,000 and 200,000 objects were propagated over time spans of one, five, ten, twenty and fifty years each. A time step of one day was chosen in each case, resulting in a combined total of approximately 11.33 billion propagation operations per tested processor. All scenarios were executed with FLORA on two different CPUs (table 5.5) and with Ikebana on three different GPUs (table 5.4). Each time, the run time of the propagations was averaged over two consecutive runs before determining the MP/s value. Memory transfers between the CPU and the GPU are not included even though they may have an impact on the overall run time; the reason for not including them here is that the amount of intermediate results that needs to be transferred back to the CPU depends largely on the use case. Long-term simulations may require only yearly output while a visualization software may need to download the data once every frame. Measurements including memory transfers are shown in section 6.4.

For FLORA, two standard Intel Core i7 CPUs were used, one in a desktop and one in a mobile environment. Both processors have multiple cores but since FLORA does not support parallel execution, only one core was used in each case. For power consumption the TDP (*Thermal Design Power*) is given; it indicates the maximum power intake for which the cooling system is designed which is usually the amount of power the device draws when running typical high-performance applications. Since the CPUs only use one core in FLORA and Ikebana is unable to push the GPUs to maximum capacity (see section 5.3.5) the actual power consumption for these tests is lower in all cases; the TDP value should therefore only be regarded as a coarse point of reference.

The graphics processors used for the Ikebana benchmarks are listed in table 5.4. Since Ikebana is written in CUDA, only processors made by NVIDIA are currently supported. The Tesla K20c is a professional level adapter that was designed specifically for GPU computing. It has more cores and faster double precision support than standard consumer level GPUs and it can be installed into high performance clusters. It also has more RAM than other graphics cards of the time with support for error correction (ECC). The GeForce GTX 860m is a mobile GPU. Since it is not available separately, table 5.4 lists no price tag. It is usually built into laptops focused on gaming and multimedia applications which come at a retail price of around 1,000 €. The processor's design for power efficiency results in a relatively low core count. The GeForce GTX 960 is a current generation consumer graphics adapter for desktop PCs that is easily availabe at retail stores. Of the current lineup, it is the card that is balanced for highest cost-effectiveness. While it has support for the faster PCIe 3.0 bus, the test system only supported the slower PCIe 2.0 variant so the card's full memory bandwidth could not be utilized.

	Tesla K20c	**GeForce GTX 860m**	**GeForce GTX 960**
Build Year	2012	2013	2014
Architecture	Kepler	Maxwell (1st Gen.)	Maxwell (2nd Gen.)
Clock Rate	758 MHz	1029 MHz	1127 MHz
Bus Interface	PCIe 2.0	PCIe 3.0	PCIe 2.0
Kernel Count	2496	640	1024
Power Consumption (TDP)	225W	ca. 100W	125W
approx. Retail Price	3,100€	(not sold separately)	200€

Table 5.4.: GPU specifications

	Intel Core i7-3820	Intel Core i7-4710HQ
Environment	Desktop	Laptop
Build Year	2012	2013
max. Clock Rate	3.6 GHz	3.5 GHz
Power Consumption (TDP)	130W	47W
approx. Retail Price [€]	350€	320€

Table 5.5.: CPU specifications

5.3.4. Performance Results

Tables 5.6 to 5.10, as well as figures 5.29 and 5.30, show the performance of FLORA and Ikebana on the different hardware setups. FLORA ran at relatively similar speeds on both processors; with the desktop CPU achieving around 0.13 MP/s and the laptop performing at around 0.1 MP/s, the desktop showed an advantage of around 30 per cent. Only little variation between the scenarios was observed; the higher number of decaying objects in longer scenarios explains the slight increase of speed compared to the shorter ones.

For Ikebana, all GPUs were able to significantly outperform both CPUs. The massively parallel architecture of the graphics processors can only be fully utilized with a large number of objects; therefore, the scenarios with 1,000 and 10,000 objects run at slower speeds. Also, the relative overhead from the non-parallel portions of the perturbation modules is larger for small object numbers. Between 100,000 and 200,000 objects, the MP/s value only shows a very minor increase implying that the algorithms maximum hardware utilization has been reached. However, as outlined in the next section, this does not mean that the hardware is running at 100 per cent capacity. At 200,000 objects, the Tesla runs at approximately 2.8 MP/s, outperforming the CPUs by factors of around 22 and 28, respectively. Each newer hardware generation raises these factors by a significant amount, with around 38/47 for the laptop GPU and and 60/75 for the newest generation hardware. The slight increase caused by the higher number of decayed objects in the longer running scenarios is observable on the GPU as well.

	N_{Obj}				
$\frac{N_P}{t_{total}}$	**1000**	**10,000**	**50,000**	**100,000**	**200,000**
1 year	0.11945	0.119712	0.119452	0.119884	0.119834
5 years	0.119956	0.120749	0.120591	0.120982	0.1209
10 years	0.120608	0.121818	0.121499	0.121888	0.121376
20 years	0.123131	0.124555	0.124283	0.124415	0.123985
50 years	0.128331	0.129064	0.129029	0.129724	0.129694

Table 5.6.: Performance of FLORA on Intel(R) Core(TM) i7-3820 CPU [MP/s]

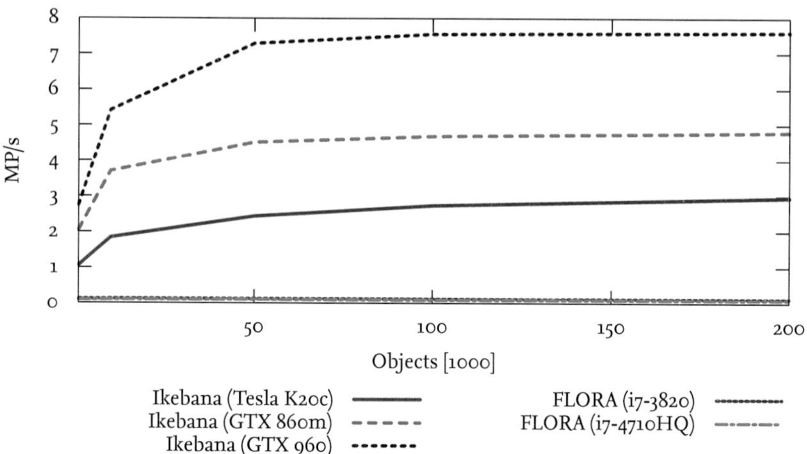

Figure 5.29.: Performance in megapropagations per second of FLORA and Ikebana on various platforms.

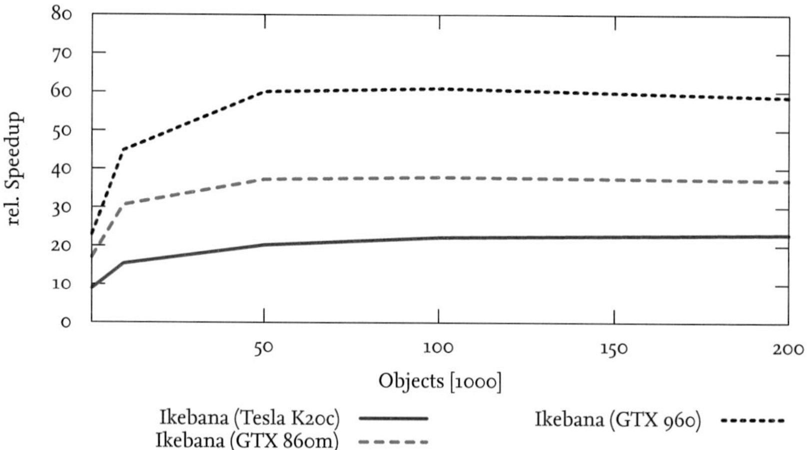

Figure 5.30.: Relative speedup of Ikebana on various platforms compared to FLORA (i7-3820).

5.3.5. CUDA Runtime Analysis

The CUDA programming language looks and feels very similar to the C language, hiding the fact that the architecture of a GPU is vastly different from that of a CPU. While recent programming languages and techniques abstract from the underlying hardware, CUDA pro-

	N_{Obj}				
$\frac{N_P}{t_{total}}$	**1000**	**10,000**	**50,000**	**100,000**	**200,000**
1 year	0.0879236	0.101952	0.094835	0.0875917	0.0991498
5 years	0.0884989	0.0997737	0.0887108	0.0885253	0.104466
10 years	0.0888536	0.093495	0.0930509	0.0891623	0.103896
20 years	0.0907781	0.0974876	0.102065	0.0911142	0.100514
50 years	0.0943272	0.103678	0.108652	0.0949654	0.09600

Table 5.7.: Performance of FLORA on Intel(R) Core(TM) i7-4710HQ CPU [MP/s]

	N_{Obj}				
$\frac{N_P}{t_{total}}$	**1000**	**10,000**	**50,000**	**100,000**	**200,000**
1 year	1.05587	1.80505	2.37784	2.66331	2.79332
5 years	1.05023	1.78512	2.36734	2.67464	2.83257
10 years	1.04503	1.79503	2.39443	2.67615	2.85134
20 years	1.04946	1.80277	2.4206	2.7535	2.81937
50 years	1.0611	1.84529	2.43957	2.74108	2.9648

Table 5.8.: Performance of Ikebana on NVIDIA Tesla K20c [MP/s]

grammers need to take this into account to achieve the best performance. While Ikebana already outperforms FLORA, a deeper analysis outlined in this section shows that there is still room for improvement. Figure 5.31 shows the run time portions of the individual perturbation models in CUDA. On all devices, the atmospherical model is by far the most load-intensive module although the actual percentages vary greatly. On the Tesla card, calculations for the atmosphere take up more than 90 per cent of the compute time while the same operations performed on the two GeForce cards use only around 60 per cent. Also, the relatively low overall performace of the Tesla comes as a surprise. Since the card has almost four times as many cores as the laptop GPU a much better performance could be expected. An analysis of the atmosphere kernel with CUDA's visual profiler (Appendix A) shows that the Tesla card's performance is limited by memory constraints while the main bottleneck of the GeForce cards is the available register count. The exact reason for this particular outcome can only be found in the optimization routines of the CUDA compiler which are subject to NVIDIA's trade secrets. Although not the most significant issue on the Tesla, the limited register count is a bottleneck common to all three cards. Since all threads in a block share the available registers, the amount of concurrent kernels is limited by the register memory that each thread requires. This means that although the hardware supports more than 2400 parallel threads, the atmospherical model requires too much memory to be able to run all of them at the same time. For this reason, the atmospherical model of Ikebana is already configured to run at lower block size of 64 threads per block. On the GTX 960, the CUDA compiler's optimization routine decided to reduce the number of occupied registers to 63 per thread compared to 76 on the other cards. This allows the GTX 960 to run 16 concurrent blocks (i.e. 1024 concurrent threads) instead of 12 blocks (i.e. 732 threads) by sacrificing faster memory access. A possible reason for this decision is the faster clock rate of the GTX 960 that allows the threads to finish in a shorter time. Neither of the cards achieve full use of all

	N_{Obj}				
$\frac{N_P}{t_{total}}$	**1000**	**10,000**	**50,000**	**100,000**	**200,000**
1 year	1.92289	3.63601	4.42572	4.58937	4.67917
5 years	2.03542	3.63923	4.44351	4.60205	4.69514
10 years	2.03567	3.65581	4.45251	4.61073	4.70606
20 years	2.03913	3.67133	4.47067	4.63292	4.73661
50 years	2.04413	3.70397	4.51904	4.69204	4.8018

Table 5.9.: Performance of Ikebana on NVIDIA GeForce GTX 860M [MP/s]

	N_{Obj}				
$\frac{N_P}{t_{total}}$	**1000**	**10,000**	**50,000**	**100,000**	**200,000**
1 year	2.40742	5.16932	7.18636	7.46922	7.5274
5 years	2.72049	5.20699	7.22574	7.47147	7.52225
10 years	2.72968	5.26549	7.22861	7.47152	7.52344
20 years	2.74429	5.32267	7.25659	7.49771	7.54349
50 years	2.74599	5.41565	7.28331	7.54792	7.5971

Table 5.10.: Performance of Ikebana on NVIDIA GeForce GTX 960 [MP/s]

available cores; however, this glimpse into the compiler's optimization process suggests that one hundred per cent occupancy does not necessarily yield the optimal performance.

For the GeForce cards, the analysis lists the kernel's relatively high use of double precision variables as an additional bottleneck. The Tesla's hardware is optimized for faster double precision calculations and therefore shows no significant slowdown. A possible reason for this is that on CUDA hardware, double precision operations are performed by linking two single precision FPUs; on the Tesla, the relatively low occupancy increases the availability of unused FPUs. For neither card the bus speed proves to be a limiting factor; therefore connecting the GTX 960 to a faster PCIe 3.0 bus cannot be expected to increase the performance of the atmospherical model.

The profiler results point at additional possibilities for optimization, some of which can be applied to all three GPUs. Concurrency could be enhanced further by limiting the amount of registers required by each thread; but with complex models such as the atmosphere, this is not a trivial task. Another example is the strong diversion at the *if*-statements that determine whether or not an object is affected by the atmosphere, and whether the equations for near-circular or eccentric orbits have to be applied. Since the SIMD architecture demands that all kernels follow the same branches, concurrently propagating e.g. a LEO and a GEO object in the same block will cause stalling. This can be prevented by sorting the objects so that only those likely to follow the same branches will be assigned to the same block. Provided that a sorting algorithm can be found with a run time shorter than the time lost by stalling, this will further enhance performance. A function for this could be provided by OPI so plugin writers would have an extra benefit without further effort. Sorting could easily take place on the CPU and would not have to be executed for each time step since only a small percentage of objects cross the border into the atmosphere's area of effect or change from eccentric to near-circular orbits. A sorting algorithm optimized for arrays with small changes should be able to perform this task very efficiently. With a divergence of nearly 100 per cent, performance can be expected to nearly double in the best case.

The analysis also shows that the atmospheric module makes no use of the relatively fast shared memory. Values that are common to all threads of a block could be placed into shared memory to increase performance. Possible candidates for this are the date-dependent values which are the same for all objects, such as the pointers into the lookup table for the atmospherical data. The constant memory would be well-suited for read-only data such as the atmospheric model's lookup table itself; unfortunately, with only 64 kilobytes available it is much too small to hold the approximately 50 megabytes of atmospherical data.

Another area of optimization is the warp size: All three cards only execute two warps per block. In the event of a memory stall, CUDA hardware can switch between warps so another set of threads can be computed while another one waits for the memory transfer to finish. If only two warps are available to choose from, the choice is limited so the probability of all threads having to wait is higher. Increasing the warp size might decrease stall times; however, since memory bandwidth is not as big an issue as register count of divergence, the expected speedup is smaller.

The final possible optimization is specific to the different bottlenecks of the Tesla GPU. The memory throughput of that card can be improved by using individual variables instead of structs such as *OPI::Orbit* and *OPI::Properties* which would greatly reduce readability. Another way to achieve this would be optimizing the structs' memory alignment with precompiler directives such as *#pragma pack*. In OPI, this would require some extra effort to ensure Fortran compatibility. Both options were ultimately discarded because they provide only minor improvements to the Tesla's performance and none to the GeForce GPUs which are not limited by memory bandwidth.

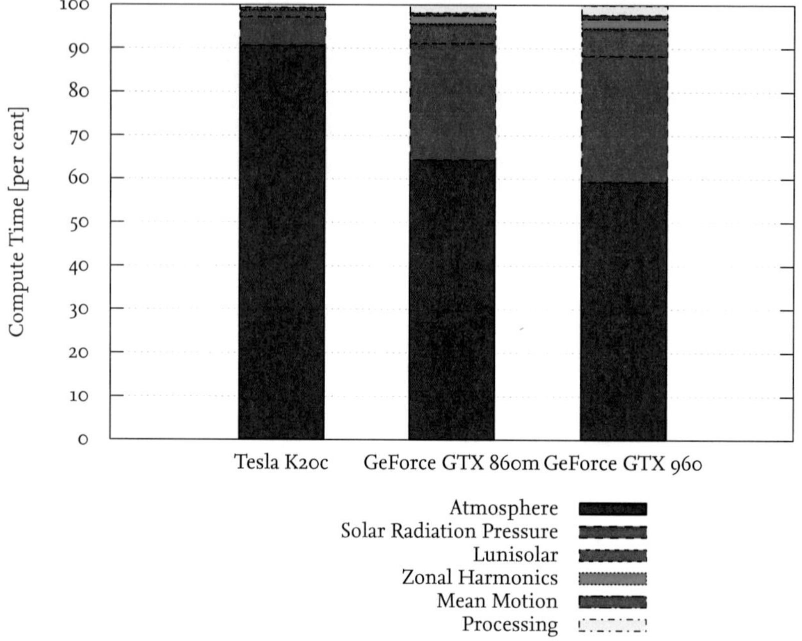

Figure 5.31.: Percentage of the total run time of each perturbation module.

5.4. Double Precision Comparison

As concluded from section 5.2, the accuracy of Ikebana is deemed sufficient for its designated use case; the continuous use of double precision variables is not necessary. However, small deviations can be observed for most objects. To determine if these issues are caused by Ikebana's lower floating point precision, the application was fitted with the new variable type *ikfp_t* which was used for all internal floating point variables. Depending on the desired precision, *ikfp_t* can be set to either *float* or *double*. The data types provided by OPI were left unchanged so the resulting orbits are still converted to single precision after each module finishes its calculation. Judging by the GPUs' specifications, using double instead of single precision leads to significant performance losses in CUDA. [NVIDIA Corporation, 2013] lists the maximum peak performances of the Tesla card at 3.52 Tflops (i.e. trillion floating point operations per second) for single precision and 1.17 Tflops for double precision. Since the Tesla cards are specifically optimized for scientific applications, regular consumer GPUs show even bigger performance gaps. To verify this, the scenarios from section 5.3.3 were executed again with *ikfp_t* set to double precision.

The performance results are listed in tables 5.11 to 5.13. While the GTX 860m and the GTX 960 show significant losses by factors of around 3.5 and 2.5, respectively, the Tesla K20c performs at almost the same speed as with single precision variables. Given its specific purpose

$\frac{N_P}{t_{total}}$	N_{Obj} 1000	10,000	50,000	100,000	200,000
1 year	0.841087	1.78385	2.31649	2.53422	2.62704
5 years	0.835525	1.80175	2.32506	2.50202	2.59523
10 years	0.833607	1.79742	2.32413	2.50324	2.64006
20 years	0.833681	1.82396	2.36799	2.53014	2.6106
50 years	0.834151	1.838	2.37197	2.55913	2.7054

Table 5.11.: Performance of Ikebana (Double Precision) on NVIDIA Tesla K20c [MP/s]

$\frac{N_P}{t_{total}}$	N_{Obj} 1000	10,000	50,000	100,000	200,000
1 year	0.834616	1.24769	1.31528	1.34153	1.3506
5 years	0.864329	1.25345	1.31753	1.34356	1.35244
10 years	0.864029	1.25527	1.31866	1.34399	1.35362
20 years	0.864404	1.25908	1.32189	1.34719	1.35699
50 years	0.866791	1.26627	1.32836	1.35402	1.3640

Table 5.12.: Performance of Ikebana (Double Precision) on NVIDIA GeForce GTX 860M [MP/s]

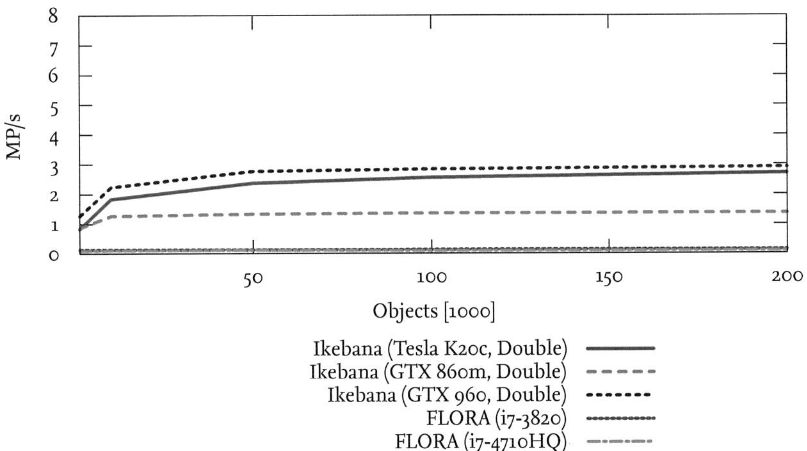

Figure 5.32.: Performance of FLORA and Ikebana with double precision: All platforms but the Tesla suffer significant performance losses.

it is expected to show the smallest performance drop; however, contrary to the specification which lists a lower peak performance for doubles, the test result almost exactly matches that

	N_{Obj}				
$\frac{N_P}{t_{total}}$	**1000**	**10,000**	**50,000**	**100,000**	**200,000**
1 year	1.20366	2.20059	2.75035	2.85016	2.88783
5 years	1.27224	2.21461	2.7424	2.85219	2.88677
10 years	1.26674	2.22022	2.76192	2.84734	2.8884
20 years	1.2703	2.23497	2.67323	2.85242	2.8927
50 years	1.26979	2.23551	2.7686	2.83752	2.9018

Table 5.13.: Performance of Ikebana (Double Precision) on NVIDIA GeForce GTX 960 [MP/s]

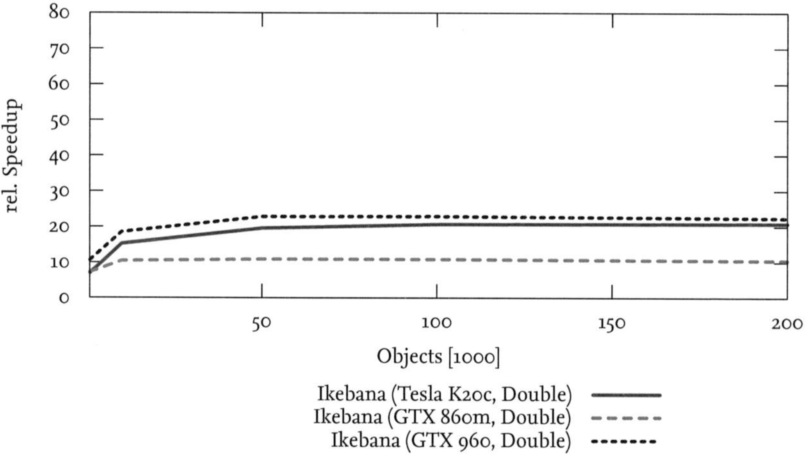

Figure 5.33.: Relative speedup of Ikebana with double precision compared to FLORA.

of the single precision version. As explained in the previous section, it is assumed that the card's peak performance is not reached and therefore no difference is visible.

Looking at the propagation results, the double precision version of Ikebana shows little deviation from the single precision variant. In the LEO and GEO regimes, the output of the original version is reproduced exactly for the vast majority of objects (figures 5.34 and 5.35); the mean decay rates of both versions are identical. In a few cases where FLORA and Ikebana deviate due to very low eccentricities, the double precision version resembles FLORA's output more closely; figures 5.36 and 5.37 show different examples of this effect. Due to the overall very small effect of the solar radiation pressure module it is likely that floating point inaccuracies resulting from it are too small to be noticed in these results. No further analysis has been conducted as this module was found to be in an incomplete state.

It should be noted that the double precision version of Ikebana has not been verified for correctness so these results should be regarded as preliminary. However, it can already be concluded that most of the floating point inaccuracies between FLORA and Ikebana are caused by the single precision OPI types. This is plausible since the OPI types are prominently used throughout the propagator. However, for the same reason, changing them to double precision is likely to cause another substantial performance drop on the consumer GPUs. With the exception of the rare low eccentricity problem, all other occurrences of floating point inaccuracies appear to have been identified and properly addressed.

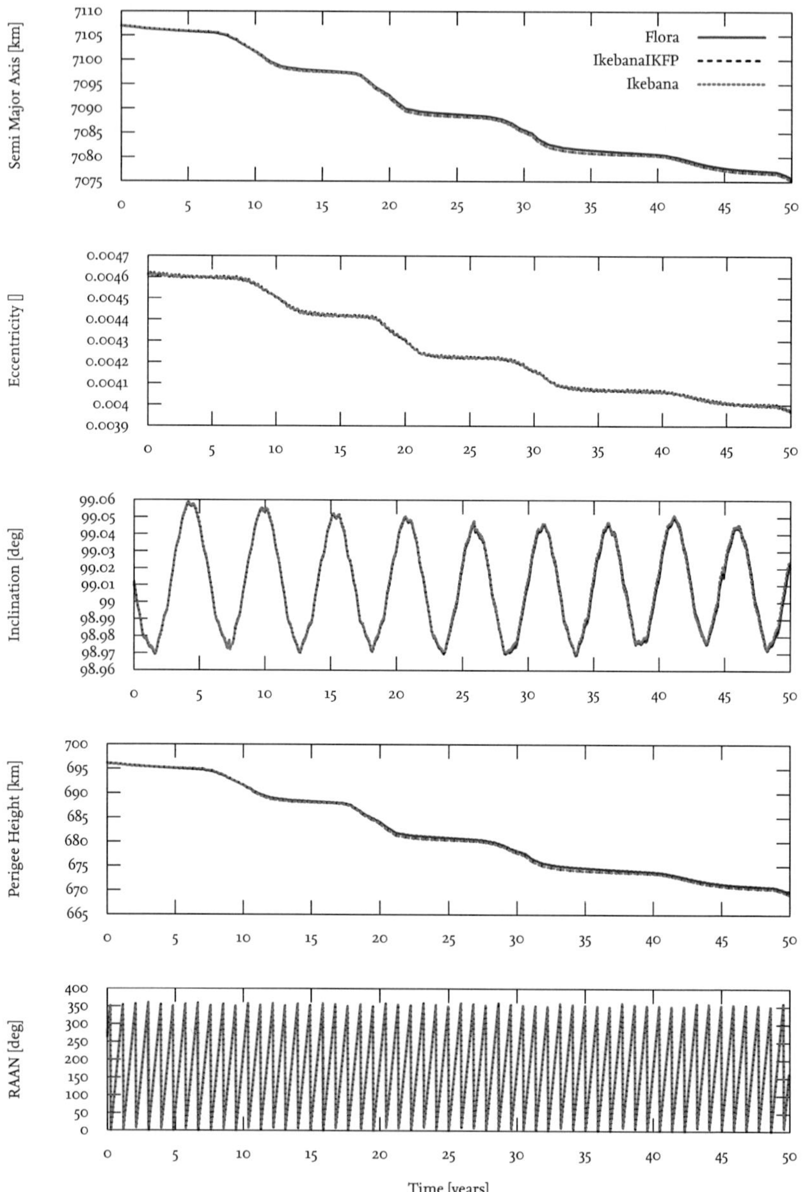

Figure 5.34.: Ikebana with double precision: Most objects show no significant deviation from the single precision version (LEO). Object number: 37452

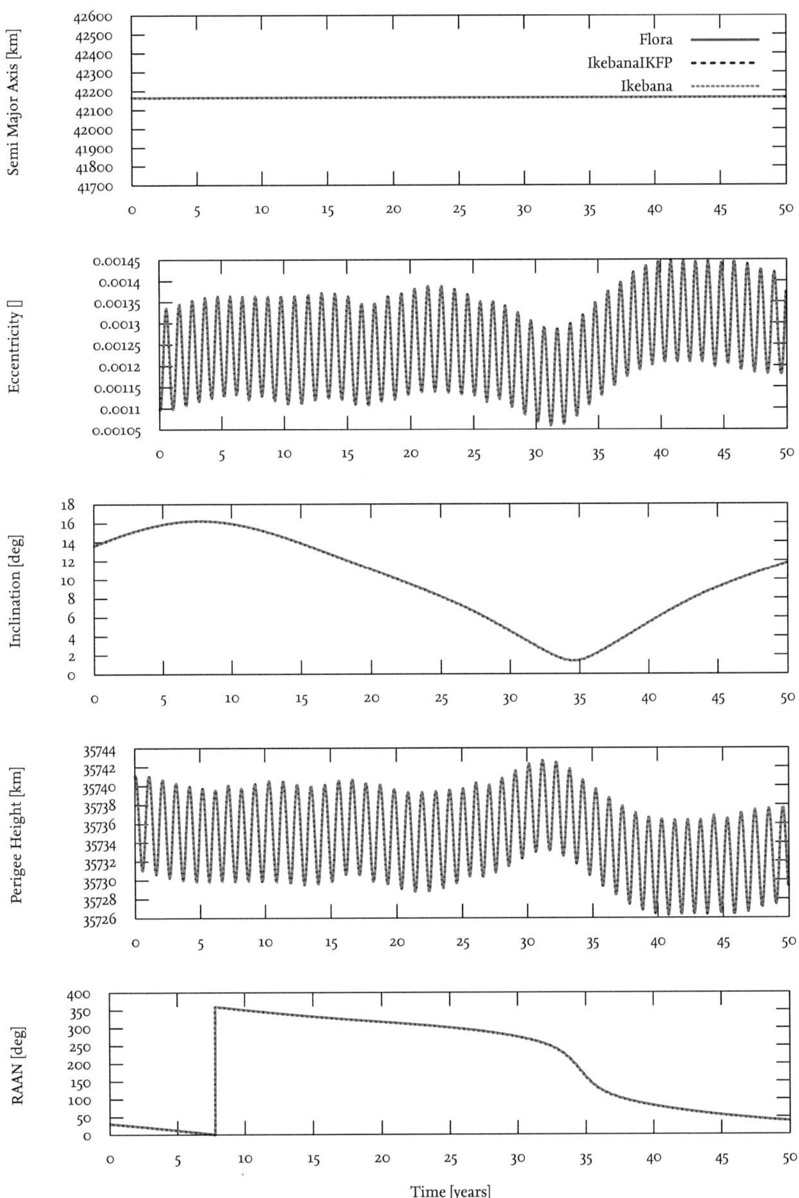

Figure 5.35.: Ikebana with double precision: Most objects show no significant deviation from the single precision version (GEO). Object number: 22963

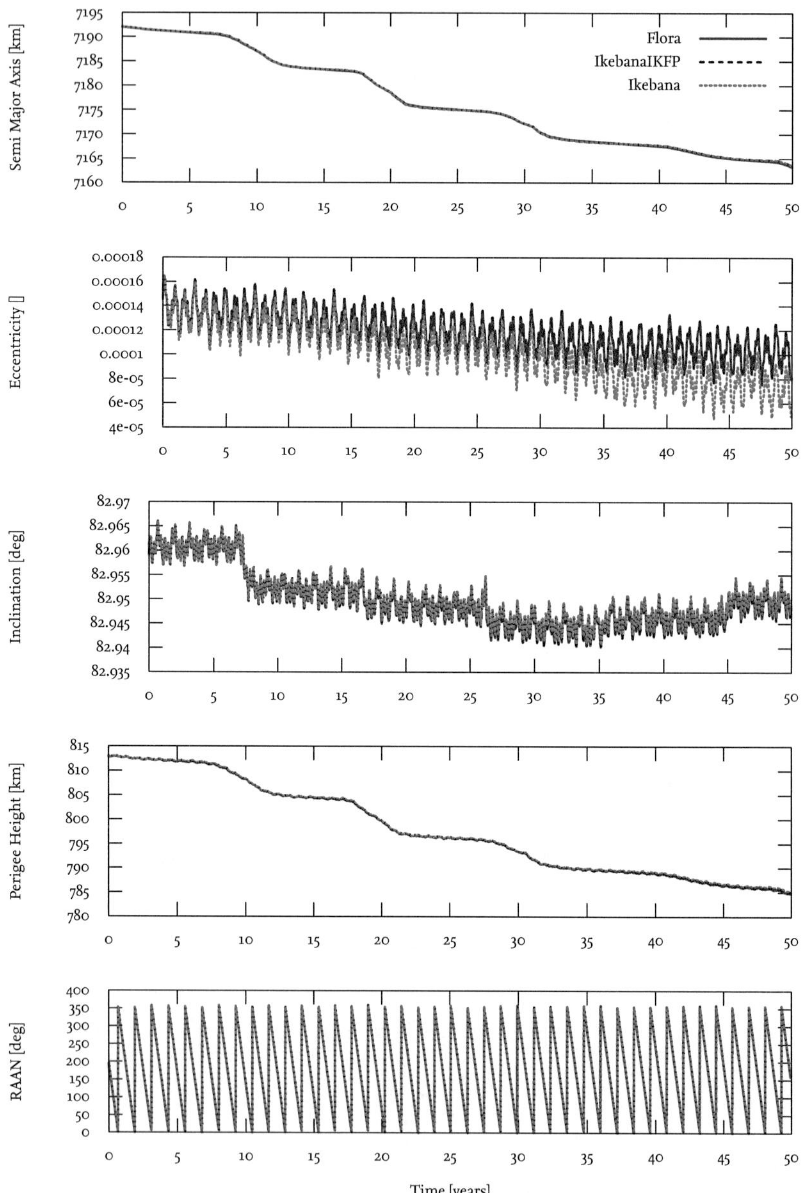

Figure 5.36.: Ikebana with double precision: In other cases, the results from FLORA are reproduced more closely. Object number: 12952

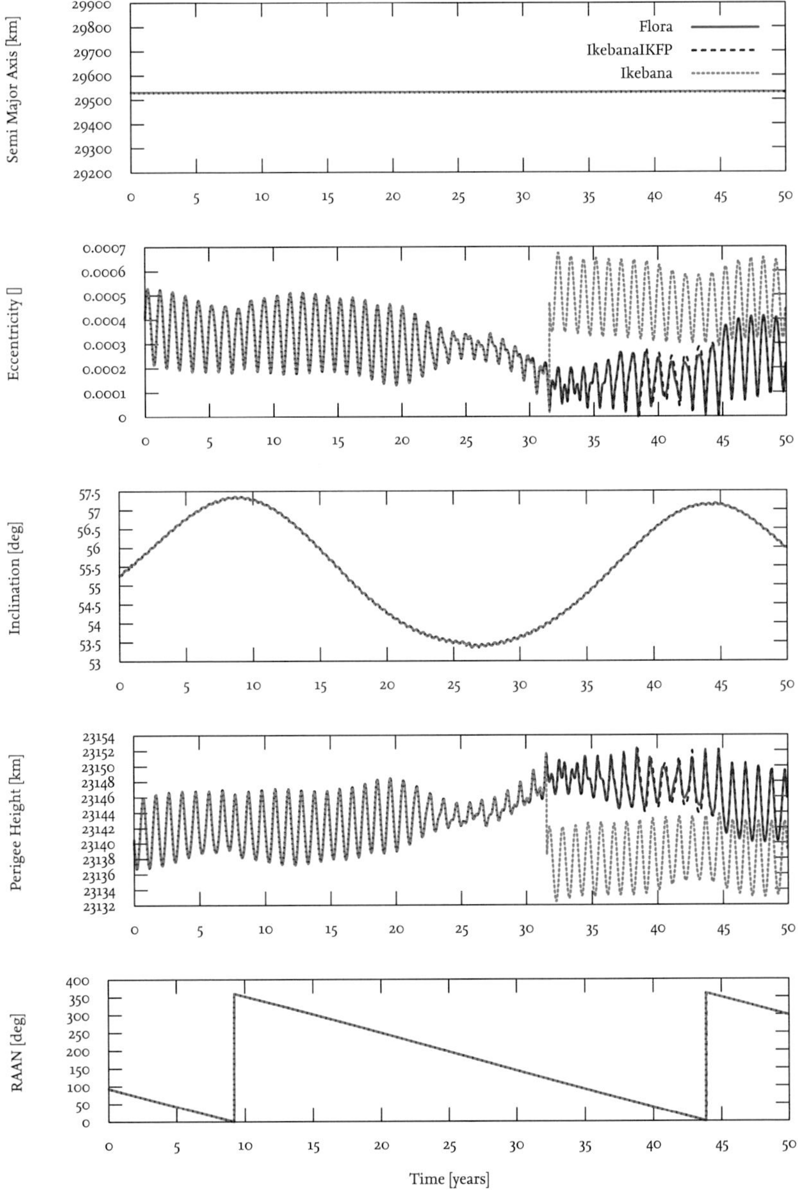

Figure 5.37.: Ikebana with double precision: The low eccentricity problem of Ikebana can be addressed by using double precision (GTO). Object number: 37847

5.5. Summary

Overall, Ikebana provides propagation results that are very close to that of FLORA. Even though some of the deviations do increase with time, all of them are well within the acceptable range for the use cases of long-term population analysis and visualization. An issue that should be observed is the artificial increase of very low eccentricities that causes unpredictable deviations. These happen in FLORA as well and are an inherent singularity of analytical propagation. While raising the precision does lessen the impact, occurrences of this issue are extremely rare and are therefore unlikely to affect the overall results in large populations. The solar radiation pressure module can be expected to cause more significant errors with HAMR objects and will have to be reevaluated before use with certain types of space debris such as MLI foil. Apart from this uncertainty, it can be concluded that for analytical propagation, sufficiently accurate results can be produced without the general use of double precision variables. However, it is also evident that reducing the floating point precision causes significant additional effort. Just as shown in the research of [Fraire et al., 2013] mentioned in section 5.2.1, porting an existing algorithm directly to the GPU and simply reducing the floating point accuracy can be problematic. Possible occurrences of precision loss have to be identified and equations have to be rearranged. This work can be extremely difficult and tedious, especially with the more complex perturbation models like atmosphere and solar radiation pressure. A brief investigation into a double precision version of Ikebana (section 5.4) suggests that the most significant locations of precision loss have been identified. In many cases, such locations consisted of a single line of code that caused major deviations for a specific class of objects. Once they were found, rearranging the equation or converting elements to double precision was sufficient to solve the problem. Remaining accuracy deviations are either a result of undetected programming errors, or the lower accuracy of OPI's data types.

The run time measurements detailed in section 5.3 show that the additional effort and the minor loss of precision can be justified by the vast performance improvement delivered by GPU computing. With a speedup factor of up to 60, the time required to generate one table of measurement data for this chapter could be reduced from approximately 2.6 days to under an hour with Ikebana. In long-term analysis, this advantage can be used, for example, to improve the results by increasing the number of Monte Carlo simulations. Every new GPU generation performs significantly faster than their predecessor, allowing the speculation that there is still much room for improvement in future hardware developments. An in-depth analysis of the atmospheric model hints at optimization opportunities that could double its performance.

The different GPUs provided the same results for all calculations. In theory, the ECC memory of the Tesla card is able to automatically correct corrupted data. During the analysis however, no situation was identified where this was applicable. For the use cases of statistical long-term analysis and visualization, such occasional errors are very unlikely to have a significant effect on the final outcome. In terms of performance, enabling or disabling the error correction on the Tesla had no significant impact on Ikebana's performance. The lack of this requirement coupled with the fast deprecation and the high price of the hardware make the Tesla card a highly uneconomic choice in this case. Unless cluster computing is required for extremely large populations or complex parallel post-processing in the host program, regular consumer GPU cards are sufficiently capable of performing analytical propagation of large populations.

6 Use Case Study: Space Debris Visualization

6.1. Overview

DOCTOR (Display of Objects Circulating in Terrestrial Orbits) is a visualization program designed to illustrate the development of the space debris environment. It is capable of drawing and animating large numbers of objects. Each object can be selected individually to display additional information such as the current and original orbits, name, type and origin. A ground track can be painted in real-time and exported as a Mercator projection (figure 6.1). The main use of the application is to generate images and animations for research and educational purposes but it has also been used for publicity and even works of art ([Peus, 2013], [Najjar, 2014], [Bundeskunsthalle, 2014]). For the latter, additional features have been implemented such as support for stereoscopic displays and the ability to place the camera on top of any moving object. The application also facilitates verification of source models for space debris objects; in [Rohrbeck, 2011] for example, it was used to identify the cause for an erroneous distribution of the West Ford needle population. With the help of the OPI interface the program can be used to visualize the performance of any propagator in real-time in order to spot unexpected behaviour. However, to perform its main objective, the depiction of the whole space debris population, it requires a very fast propagator.

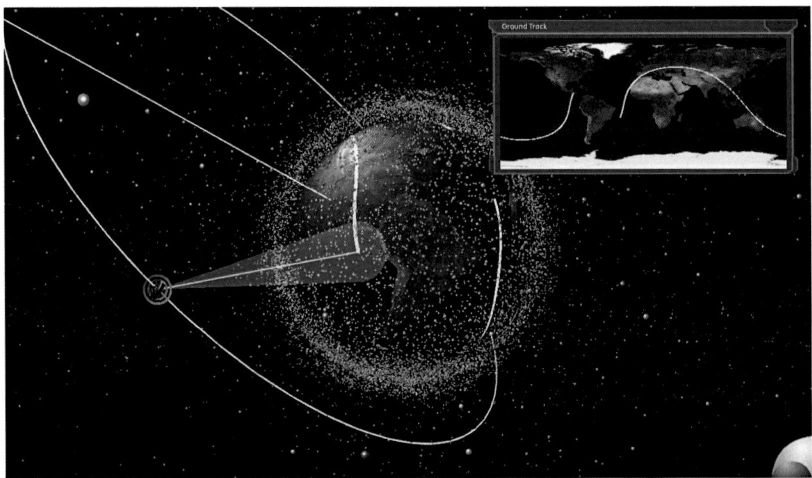

Figure 6.1.: Ground track projection in DOCTOR.

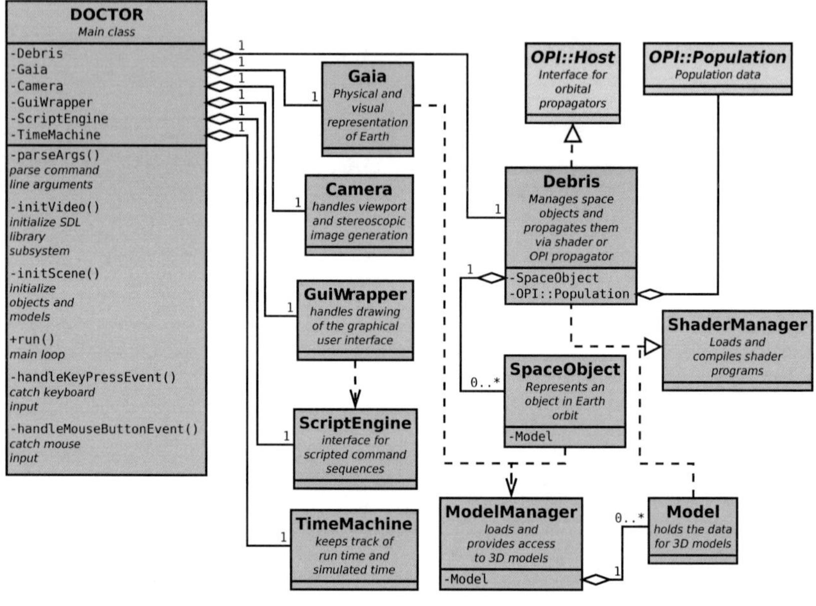

Figure 6.2.: Simplified UML diagram of DOCTOR.

6.2. Classes

Figure 6.2 shows an UML diagram of DOCTOR's most important classes which are described in detail in the following sections. DOCTOR makes use of several open source libraries which are noted in the appropriate sections.

6.2.1. DOCTOR::DOCTOR

This is the main class of the visualization. It is responsible for setting up and running the program which includes parsing command line arguments from the user and loading the population data and other input files such as models and textures. It uses the open source Simple Directmedia Layer (SDL) library ([Latinga, 2014]) to interface with the graphical environment of the operating system, provide a basic OpenGL context and process input from mouse and keyboard. At the heart of the class is the main loop which runs continually until the user issues the *quit* command (figure 6.3). It first checks for mouse movement and button presses and executes the functions associated with that specific input. For example, if a window of the graphical user inferface was clicked, the key presses are forwarded to the *DOCTOR::GuiWrapper* class which handles that input. The main loop then issues an update command to all classes that need to perform per-frame tasks, most importantly *DOCTOR::TimeMachine* which keeps track of run time and simulated time. Subsequent updates are performed by other classes such as *DOCTOR::Gaia* which calculates the Earth's rotation. Finally, the loop calls the update function on *DOCTOR::Debris* to propagate the objects and

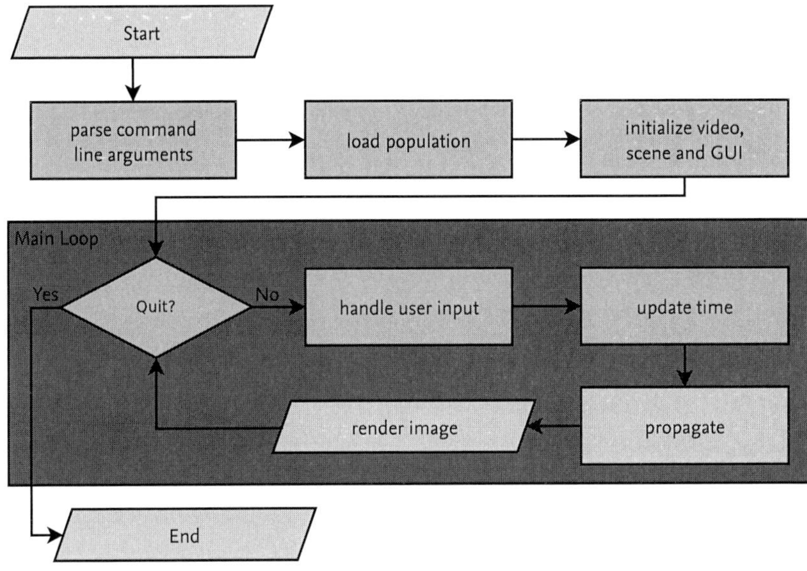

Figure 6.3.: Flowchart illustrating DOCTOR's initialization and main loop.

then executes OpenGL operations that run the shader programs and draw the frame. To provide a fluent animation, the propagation time step is chosen based on how much time has passed since the drawing of the last frame.

6.2.2. DOCTOR::SpaceObject

This class represents an object orbiting Earth. In addition to the orbital elements and properties defined by *OPI::Population* this class stores additional information fetched from the Satellite Situation Report (SSR), a document which is periodically released by NASA and can be loaded by DOCTOR on startup. The object ID that is used in DOCTOR's input files contains the object type (e.g. payload, rocket body, explosion fragment, etc.), its country of origin and the NORAD ID if it represents a catalogued object. In that case, the ID is checked against the SSR; if a match is found, additional information such as name, launch and decay dates are read from the catalogue and stored in the *SpaceObject* class. The object's color is chosen based on its type, the size based on its diameter. The orbital parameters are analyzed to determine the orbit type (e.g. LEO, GEO, Molniya, etc.). Originally, the *SpaceObject* class had a propagation function that was called on each update and each instance, thus propagating the whole population sequentially on the CPU. As this quickly proved unfeasible for populations larger than a few thousand objects this function was deprecated and replaced with the methods described in section 6.3.

6.2.3. DOCTOR::Debris

This class manages an array of *SpaceObject* instances which represent a population. Since the deprecation of sequential propagation this class is also responsible for propagation. For this purpose it implements the *OPI::Host* interface and checks for propagator plugins at startup. If a plugin is found, it uses the list of SpaceObjects to compile an *OPI::Population* from its data and calls the plugin's *propagate* function for every frame. If no plugin is found a fallback to a simple shader-based propagation method is used (see section 6.3). The class has additional functions used for providing information to the user. One function identifies the object at the position of the mouse if a left-click was registered. A pointer to the identified object is stored which can be used to display information about its current position and to draw its orbit and ground track. Another function can apply a filter to the population to display only objects of a specific type, name, size, nationality and other properties. Other functions control display-specific properties such as the objects' scale on screen or the thickness of the ground track.

6.2.4. DOCTOR::TimeMachine

This class controls date and time information. Most of its methods are static so that all other classes can easily access current information. The default date is set to January 1st, 1950, the point from which the Modified Julian Date is defined[1]. From there the time progresses with the application's run time, with a scaling factor that controls the animation speed. The default factor is 120, i.e. one second of run time corresponds to two minutes of propagation time. This value can be decreased or increased with the "<" and ">" keys. A similar function can be used to fast-forward the current time by a day, a month or a year. The class keeps record of the drawing times of the last frames which are used to set the time step for propagation. Other functions include controls such as pausing time, locking the frame rate to a fixed value, changing the default Julian date and calculating the Sun's position for lighting effects.

6.2.5. DOCTOR::GuiWrapper

This class is responsible for drawing the graphical user interface. It uses the CEGUI open source library ([Turner, 2014]) which is capable of drawing common GUI components directly in OpenGL. The window elements of the GUI are defined in a layout file that is loaded on startup. The CEGUI library handles clicks and key presses that affect the GUI elements; the functionality is implemented in the *GuiWrapper* in the form of *callback functions*, i.e. functions to which pointers are provided to the CEGUI library to be executed on a specific event. The *GuiWrapper* is accompanied by a class called *CEGUIPhysFSResourceProvider* which serves as an interface to DOCTOR's *FileManager* (see section 6.2.7); it enables the GUI to load image and layout files directly from DOCTOR's universal resource file, *doctor.dat*.

6.2.6. DOCTOR::ScriptEngine

DOCTOR's scripting engine features an embedded interpreter for the Lua scripting language ([Ierusalimschy et al., 2006]) and allows users to control DOCTOR's most important functions via Lua scripts. These include loading and deleting populations, applying filters, controlling animation speed, moving the camera on predefined paths, recording video and setting PropagatorProperties via OPI. Script files are usually provided at start via the command line and parsed in every frame; the auxiliary function *runOnce* is provided that allows the user to trigger commands at a specific point in simulation time. Alternatively, Lua commands can be

[1] Since OPI takes time information in the regular Julian date format, it is converted for propagation.

entered directly through the GUI's console window (figure 6.4). Scripts are used to execute a predefined set of input commands that can be used for automatic creation of videos such as the excerpt shown in listing 6.1.

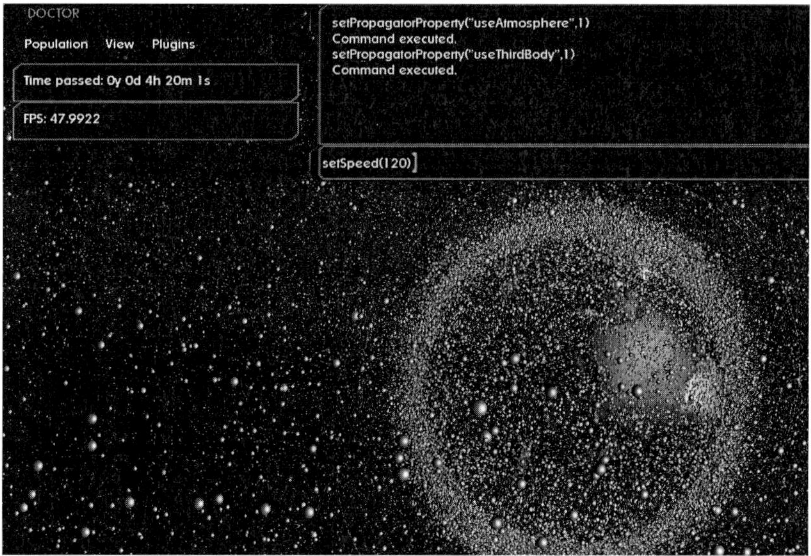

Figure 6.4.: The GUI's scripting console allows direct input of Lua functions for controlling animation and setting PropagatorProperties for OPI plugins.

Listing 6.1: Example Lua script for automating a control sequence.

```
function init()
  lockFramerate(35)                 -- set a fixed frame rate for video recording
  setSpeed(1)                       -- set speed to real time
  clearPopulation()                 -- remove all objects
  loadPopulation("clouds.sim")      -- load new object file
  setFilter("pgheight", 200, 200000)
  applyFilter()                     -- don't show objects lower than 200 km
  selectObject(99999)               -- select object with given ID
end

runOnce(0, init)                          -- run init function at second 0
runOnce(6, setSpeed, 720)                 -- increase speed after six seconds of sim time
runOnce(16373, selectObject, 99998)       -- select an object after ~4.5 hours of sim time

-- steadily increase the speed shortly afterwards
for i = 1, 10 do
  local execTime = 17000 + i * 200
  local newSpeed =   720 + i * 120
  runOnce(execTime, setSpeed, newSpeed)
end
```

6.2.7. Auxiliary Classes

In addition to the classes controlling DOCTOR's main features, the application uses several auxiliary classes that handle the underlying data. The *DOCTOR::FileManager* is the same class that is used by Ikebana; it uses the PhysicsFS library ([Gordon, 2010]) to load files from an archive called *doctor.dat*. This archive contains textures, GUI resources, 3D models, basic scripts and shader programs. All of DOCTOR's classes access their respective data through the *FileManager*. The class is also responsible for writing output files such as screenshots and ground track projections. The *DOCTOR::ModelManager* provides access to 3D models which are loaded via the Open Asset Import Library ([Gessler et al., 2014]) and represented by the *DOCTOR::Model* class. The latter stores the model data in a format that OpenGL can process as well as the shaders required for drawing the model. Shader programs are handled by the *DOCTOR::ShaderManager* class. It loads the shader code from the resource file, uses the appropriate OpenGL functions to compile them into executable GPU binaries and runs them when required. It also provides basic functions for debugging shader code and checking hardware capabilities. Finally, the *DOCTOR::TextureManager* provides access to texture images. They are converted into an OpenGL-compatible format and uploaded to the GPU upon request. Each texture is assigned a unique identifier which other classes can use to order a specific image. Textures, like models, are loaded from the resource file dynamically when they are first requested; alternatively, it is possible to load all of them at startup to prevent loading times during the application's execution.

The remaining classes are *DOCTOR::Camera* which controls the movement of the camera and parameters for stereoscopic vision, and *DOCTOR::Gaia* which is responsible for the physical and visual representation of the Earth.

6.3. Propagation

DOCTOR supports two types of propagation. The first is a GPGPU approach using shader programs written in GLSL, the second is plugin-based propagation using OPI. The two methods are outlined the following sections.

6.3.1. GPGPU Approach

The initial version of DOCTOR propagated the objects sequentially on the CPU. When this proved infeasible for smooth animation of large populations, a shader-based propagation algorithm was used. It was first published in [Möckel et al., 2011] as a proof-of-concept for the performance benefits of the GPU. The concept is shown in figure 6.5: The population is represented as a particle system consisting of a two-dimensional grid of vertices, one for each object. Two textures are loaded into GPU memory: The first contains properties such as object type and size. These are used to determine the size and color of the objects; the fragment shader then replaces each vertex with a two-dimensional image of a ball with the given color and size. Additional lighting effects are applied based on the Sun's position and predefined material properties such as shininess and opacity. The second texture contains the orbital elements; from these, the vertex shader calculates the true anomaly for each object and transforms the resulting position into Cartesian coordinates; the vertex is then placed at that point in the scene. The data is encoded into the textures in such a way that the texture coordinates map one set of data to each vertex based on its position in the two-dimensional array. In a later version, the textures are replaced with *vertex attributes*, an OpenGL data type that can be used to assign properties to vertices directly without having to use textures.

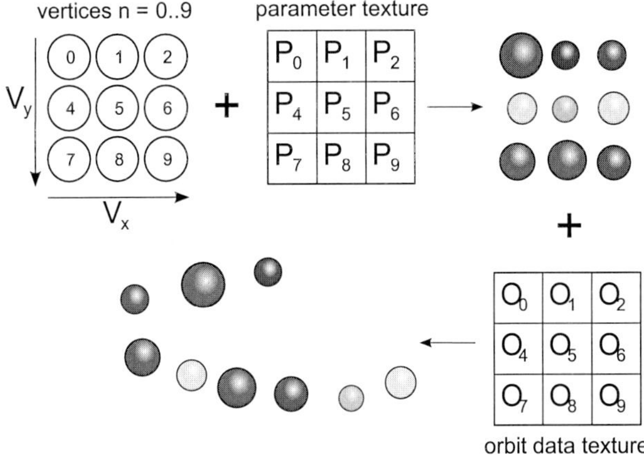

Figure 6.5.: Illustration of the GPGPU propagation from [Möckel et al., 2011]: The orbital data and parameters were provided as textures and used by a vertex shader to calculate the true anomaly.

The vertex shader discards the positional information after drawing the objects which means that there is no trivial way to use the results on the CPU. This makes tasks such as displaying the current position or drawing a perturbed orbit extremely difficult. For identifying the object which was clicked a technique called *color picking* is used. When a mouse click is registered, a special frame is rendered in which the colors of all objects are replaced by a unique color that is generated from their coordinates in the vertex array. All other objects such as the Earth, GUI and 3D models are hidden. Using the OpenGL function *glReadPixels*, the color of the pixel at the position of the mouse is returned. By using the inverse of the function used to generate the object colors, the index of the clicked object can be calculated. Since the object is picked based on the 2D screen projection of the scene this method is unable to return the 3D Cartesian coordinates of an object. For some additional features that require the object's position, the clicked object is propagated again on the CPU and the position is stored in the corresponding *SpaceObject*. This involves duplicating propagation code on the CPU which has to be kept up-to-date when the shader is changed. For this reason, the shader is kept as simple as possible; it does not calculate any perturbations except the change in the right ascension of the ascending node caused by zonal harmonics which can be added optionally. Since the inclusion of OPI, shader-based propagation is only used as a fallback for the case when no OPI propagators are found or they are unsupported on the current hardware. Due to the limitations of the shader approach many features such as drawing a ground track or displaying a 3D model at the object's position are only available when OPI is used.

6.3.2. OPI Approach

In the context of the research conducted by [Thomsen, 2013] DOCTOR was outfitted with OPI to provide a generic propagation and collision detection interface. The *DOCTOR::Debris* class was modified to inherit the methods from *OPI::Host*. Upon instantiation, it initializes OPI and checks for plugins in the given folder. If no suitable plugin is found the shader fallback is used. DOCTOR can use any OPI plugin but due to its high performance demands it is recommended to use a CUDA-capable propagator unless the population consists only of a small number of objects. In addition to being compatible with the system's hardware, the propagator requires support for generating Cartesian coordinates in order to work. By default, DOCTOR supplies a very simple CUDA-based propagator which is a port of the one used in the vertex shader. Comparing listings 6.2 and 6.3 shows that GLSL and CUDA are very similar; the only differences between the two implementations are the additional identifiers used in CUDA and the different names of the vector data types (*vec3* in GLSL versus *float3* in CUDA).

In every frame, the update function of the *DOCTOR::Debris* class calls the *propagate* function of the selected plugin using the time step provided by *DOCTOR::TimeMachine*. A shader is still used to draw the objects but instead of arranging the vertices in a grid and providing the orbital data the vertex positions are set from the Cartesian coordinates supplied by the plugin. Color and size information are provided as vertex attributes like in the original approach. To maintain a high accuracy, larger propagation time skips are subdivided into several propagation steps. If the user fast-forwards by a day, a loop is executed that runs 24 propagations with a fixed time step of 3600 seconds. Likewise, month and year skips are divided into smaller time steps. Contrary to shader propagation, the positional information for all objects in the population can be used on the CPU. It is downloaded from the GPU once every frame and stored in the respective *SpaceObjects*. This means that up-to-date information is readily available for all objects and therefore it does not need to be generated specifically for a single clicked object on the CPU. This can be used to easily implement additional features such as showing multiple perturbed orbits at once or displaying advanced population statistics. Although this additional data allows for different approaches to identifying a clicked object, the original code was kept as it proved to be very efficient.

Listing 6.2: Simple object propagation in an OpenGL vertex shader.

```
vec3 propagate(float years, float seconds,
  float sma, float ecc, float inc, float raan, float aop, float phi)
{
  float orbit_period = 2.0*PI * sqrt(powf(sma,3.0f) / RMUE);
  float t = mod(years*SECS_PER_YEAR + seconds, orbit_period);

  // calculating mean and excentric anomaly
  float mean_anomaly = mod(sqrt((RMUE * t * t) / pow(sma,3.0))+phi, 2.0*PI);
  float excentric_anomaly = mean2excentric(mean_anomaly, ecc);

  // converting excentric anomaly to true anomaly
  float sin_ea = sin(excentric_anomaly/2.0);
  float cos_ea = cos(excentric_anomaly/2.0);
  float true_anomaly = 2.0 * atan(sqrt((1.0 + ecc)/(1.0 - ecc)) * sin_ea/cos_ea);

  // based on the true anomaly, calculate Cartesian object coordinates
  float u = true_anomaly + aop;
  vec3 w = vec3(cos(u) * cos(raan) - sin(u) * sin(raan) * cos(inc),
              cos(u) * sin(raan) + sin(u) * cos(raan) * cos(inc),
              sin(u) * sin(inc));

  float p = sma * (1.0 - pow(ecc,2.0));
  float arg = 1.0 + (ecc * cos(true_anomaly));
  float r = p / EPSILON;
  if (arg > EPSILON) r = p / arg;
  return w * r;
}
```

Listing 6.3: Simple object propagation in CUDA. The code is almost identical to the vertex shader code.

```
__host__ __device__ void propagate(float years, float seconds, float sma,
  float ecc, float inc, float raan, float aop, float phi, OPI::Vector3& position)
{
  float orbit_period = 2.0f * PI * sqrt(powf(sma,3.0f) / RMUE);
  float t = fmod(fmod(years*SECS_PER_YEAR + seconds, orbit_period);

  // calculating mean and excentric anomaly
  float mean_anomaly = fmodf(sqrtf((RMUE * t * t) / powf(sma,3.0f))+phi, 2.0f*PI);
  float excentric_anomaly = mean2excentric(mean_anomaly, ecc);

  // converting excentric anomaly to true anomaly
  float sin_ea = sin(excentric_anomaly/2.0f);
  float cos_ea = cos(excentric_anomaly/2.0f);
  float true_anomaly = 2.0f * atan(sqrtf((1.0f + ecc)/(1.0f - ecc)) * sin_ea/cos_ea);

  // based on the true anomaly, calculate Cartesian object coordinates
  float u = true_anomaly + aop;
  float3 w = make_float3(cos(u) * cos(raan) - sin(u) * sin(raan) * cos(inc),
        cos(u) * sin(raan) + sin(u) * cos(raan) * cos(inc),
        sin(u) * sin(inc));

  float p = sma * (1.0f - powf(ecc,2.0f));
  float arg = 1.0f + (ecc * cos(true_anomaly));
  float r = p / EPSILON;
  if (arg > EPSILON) r = p / arg;

  position.x = w.x*r;
  position.y = w.y*r;
  position.z = w.z*r;
}
```

6.4. Performance

DOCTOR's performance was measured on the GeForce GTX 960 with various configurations listed in table 6.1. For this test, the speed is given in frames per second (FPS) for Ikebana with and without perturbations, the fallback propagation shader and the simple CUDA-based plugin that is provided as a default. Since the graphics driver limits the maximum frames per second to 60 to conserve energy, higher values are not listed. The results show that DOCTOR is capable of displaying the 1mm population of over 200,000 objects based on MASTER population data at a smooth frame rate of almost 40 frames per second when propagated with Ikebana and all perturbations enabled. As this is a different population from the one used in chapter 5 the corresponding Benchmark Index is slightly higher, around 8.1 MP/s for this configuration. Using the reference population, DOCTOR displays 150,000 objects at a frame rate of 39.2 fps and 1.5 million objects at 4.6 fps. This corresponds to around 6-7 MP/s which is a little below the 7.5 MP/s that were measured for this configuration. This can be explained by the fact that the GPU, in addition to performing the propagations, is also busy with drawing the scene in OpenGL. Also, DOCTOR does not yet have support for interconnecting CUDA and OpenGL; therefore, instead of drawing the objects directly from CUDA memory as supported by those libraries, they are transferred to the CPU in every frame and uploaded to the GPU again. The shader propagation performs fastest because there is no transfer of data between CPU and GPU memory; the CUDA plugin derived from it is slowed

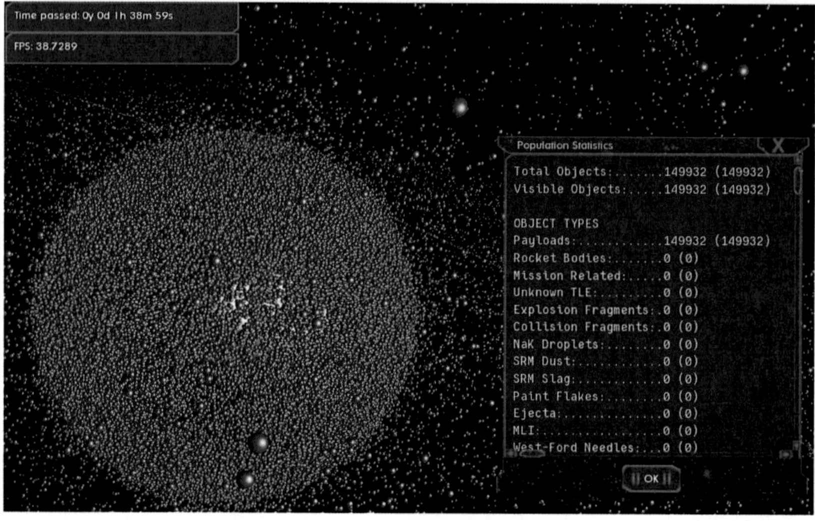

Figure 6.6.: Ikebana propagating 150,000 objects in DOCTOR at almost 40 frames per second.

down by this transfer[2]. At approximately 4 million objects, Ikebana drops to 5.5 frames per second, even with no perturbations enabled, while the shader still performs at 30 fps and the CUDA plugin at around 11 fps. This discrepancy is caused by the relative overhead of Ikebana which returns more information and manages additional data structures such as the delta orbit and *PropagatorProperties*.

Population	Objects	Ikebana (all pert.)	Ikebana (no pert.)	CUDA	Shader
10 cm	**23,000**	60 fps	60 fps	60 fps	60 fps
1 cm	**110,000**	60 fps	60 fps	60 fps	60 fps
1 mm	**219,000**	37.1 fps	60 fps	60 fps	60 fps
0.1 mm	**1,500,000**	7.2 fps	13.9 fps	26.6 fps	60 fps
Ejecta	**3,900,000**	2.8 fps	5.5 fps	10.6 fps	29.9 fps
Reference	**150,000**	39.2 fps	60 fps	60 fps	60 fps
Reference	**1,490,000**	4.6 fps	14.1 fps	26.8 fps	60 fps

Table 6.1.: Speed of DOCTOR with Ikebana, simple CUDA and shader propagation (in frames per second)

[2]The performance plot shown in chapter 5, figure 3.1 does not include the data transfer which is why CUDA performs faster there.

7 Conclusions and Further Research

7.1. OPI

The Orbital Propagation Interface described in chapter 3 was designed to facilitate the integration of orbital propagators into existing software applications. The implementation created by Patrick Thomsen was used in a variety of different software tools used in the context of this work: FLORA, Ikebana and the simple CUDA propagator used in DOCTOR were all implemented as OPI plugins. DOCTOR itself serves as an OPI host, as does the Ikebana frontend and the tool that was used to compare the propagators. Both FLORA and Ikebana could seamlessly be transferred from one host application to another without further concern. OPI is currently being used in ongoing research projects at the Institute of Space Systems to decouple the development of the host application from the propagator and to be able to easily update the propagation algorithm at a later state.

OPI was released as open source software with the goal to spread it and encourage further development by other researchers with different use cases. If adopted by others it has the potential to simplify the exchange of propagation algorithms and individual perturbation models between research groups. By releasing their work as an OPI module they would allow others to directly integrate and use it without having to port or adapt it. Modules could even be distributed and licensed in closed source form although this would contradict the spirit of open science. In section 3.3, implementation guidelines for orbital propagators as well as host applications were introduced. These are intended to serve as a starting point for further discussion and improvement of the interface.

Being in a very early state of development, many feature enhancements for OPI are conceivable for the future. The most obvious is the support for other GPU computing languages. This step has already been prepared by moving all CUDA code to a separate module. To complement orbital propagators another type of plugin, the *PopulationModifier*, would be a valuable addition. As the name states, it would allow to apply modifications to a population based on a given set of parameters. A corresponding class could be derived from *OPI::Module* as shown in figure'7.1. The interface would be similar to *OPI::PerturbationModule* described in section 3.2.5, except that instead of delta orbits, the output consists of the modified population.

Such a plugin could be used in long-term analysis to simulate satellite launches, debris mitigation measures, active removals or the generation of explosion and collision fragments. An example is shown in figure 7.2: The *PopulationModifier* takes as input an *OPI::Population* and the current date via its standard interface; additional information such as launch traffic data can be provided via an input file or via the property-related functions inherited from *OPI::Module*. The resulting population would be the same as the original with added objects that were launched at the given time. A *PopulationModifier* could also be used to simulate collisions: Using input from a collision probability plugin as outlined in section 3.2.9, it could replace collided objects of the population with a debris cloud generated by a breakup model.

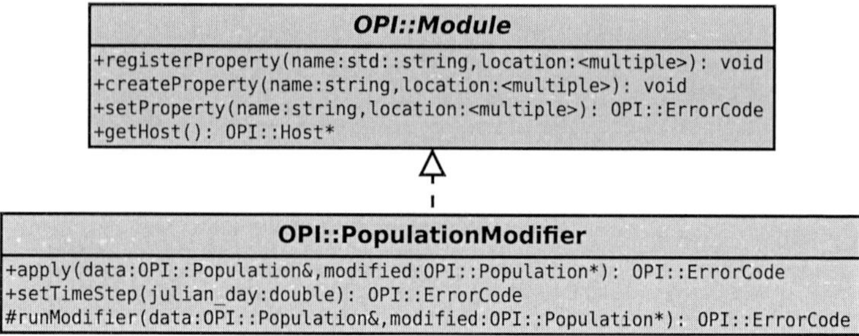

Figure 7.1.: UML diagram of the proposed *OPI::PopulationModifier* plugin type.

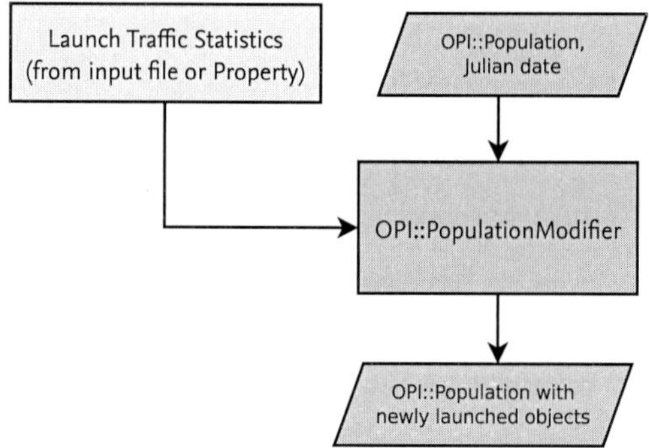

Figure 7.2.: Example of an *OPI::PopulationModifier* used to add launch traffic to a population.

7.2. Ikebana

The accuracy analysis in section 5.2.4 shows that Ikebana delivers good results compared to FLORA. For determining the absolute accuracy, one test against TLE data was performed in section 5.2.5; FLORA was validated against a numerical propagator. All investigations show that both propagators are suitable for long-term analysis. Although Ikebana is slightly less accurate than FLORA, the deviations caused by its lower floating point precision are in an acceptable range. While the TLE data comparison looks very promising, further tests are desirable to reinforce this conclusion.

The most significant difference between FLORA and Ikebana lies in the atmospheric model which causes the semi major axis of many object to decline slightly faster in Ikebana. This leads to a decay rate that is modestly higher as shown in table 5.3. While it is assumed that the divergence is small enough to be statistically insignificant in a long-term analysis, this has not yet been verified empirically. A successor to the project described in [Möckel et al., 2015] which includes further studies on the effect of debris removal and mitigation measures on the space debris population is currently in preparation. It is intended to replace FLORA with Ikebana for these long-term simulations and validate the propagator in a practical context.

Arguably the most satifying result of this work is its immediate practical value. Using GPU-computing techniques, Ikebana's performance was increased by a factor of up to 60 compared to FLORA run on a modern 6th generation Intel Core i7 CPU. Not only does this improvement justify the accuracy trade-off for long-term studies, it also enables the propagator to be utilized for interactive object visualization. Being able to use the same propagator for both scenarios has a number of significant advantages: From a software developer's point of view, it prevents unnecessary code duplication; from a researcher's stand point, it allows the validation of the algorithm in a different context. Some errors that remain undetected in one use case might become immediately apparent in another.

The different architecture of the GPU requires different optimization techniques than regular processors, even when developing for a multicore CPU. Simply converting the code from one language to another is not enough to take full advantage of the platform's capabilities and to avoid the pitfalls of its shortcomings. The analysis of the CUDA profiler results in section 5.3.5 hints at further opportunities for improvement. Implementing these is likely to provide another performance boost for Ikebana.

8 Outlook

8.1. GPU Computing

Computer science has a history of advancing at a quick pace. With that in mind, Mr Carmack's statement quoted in chapter 1 can easily be misinterpreted to imply that software developers merely have to wait for new hardware to become available. However, the paradigm shift towards massively parallel computing architectures presents a major change in the way algorithms have to be designed, and this change is only beginning to trickle into the mainstream. But even without a new major paradigm shift, existing software must be upgraded regularly because performance trade-offs may shift with hardware updates. As an example from graphics programming, [van Waveren, 2013] shows the changes that are necessary to update a 3D graphics engine for modern computers after about ten years because the performance balance between CPU, GPU and memory has shifted, deprecating old performance tweaks and introducing new bottlenecks.

The advances in parallel computing, in terms of both hardware and software, are in continuous motion. During the years in which the research presented herein was conducted, several new versions of CUDA have been released, each reducing the amount of manual optimization that was necessary to achieve the best results. It can be expected that future versios of CUDA, OpenCL and newly emerging frameworks like OpenACC will continue this trend of hiding the hardware level from users. All of these changes will simplify the development of new algorithms. OPI can profit from this in the future because its automated memory management routines can be simplified. However, given the amount of manual optimization on the memory and register level that is still required in Ikebana it is difficult to forsee the extent to which automated optimizations will be able to deliver optimum performance.

8.2. Numerical Propagation

So far this work has covered analytical propagation; OPI has been designed based on this premise. The interface should be fit for semi-analytical propagation as well. For numerical propagators, the interface as well as the GPU computing techniques used for Ikebana will have to be revised. Some numerical propagators take object data as a set of Cartesian coordinates representing a position in one of several coordinate systems. Applications like ZUNIEM are able to convert from Keplerian orbits but for use cases with very high demands for precision, the resulting loss of information might not be acceptable. Furthermore, different conversion algorithms have to be used depending on whether the host delivers single-mean, double-mean or osculating elements which would all have to be added to the interface.

Regarding GPU computing, numerical propagation is challenging as well. Numerical algorithms rely on an integrator that adjusts its step sizes dynamically. Depending on the shape of the orbit and the object's mean anomaly, step sizes may differ greatly for individual objects. This contradicts the SIMD principle that GPU hardware is based on, which delivers optimal results only if the kernel follows the same instructions for every object. Divergence can be counteracted by dynamically grouping the objects based on their orbit types, positions and predicted step sizes; the effectiveness of such measures will have to be studied further. The floating point deviations present in Ikebana are acceptable given the propagator's overall accuracy. For a high-precision numerical propagator, the impact of such errors on the results will be much larger. Whether they can be kept within acceptable limits needs to be researched.

A CUDA Profiler Report for the Atmospherical Model

A.1. GeForce GTX 860m

Analysis Report

atmosphereGPU(OPI::Orbit*, OPI::ObjectProperties*, OPI::Orbit*, int, tAtmoData*, int*, double*, float, float)

Duration	6.796 ms (6,795,909 ns)
Grid Size	[782,1,1]
Block Size	[64,1,1]
Registers/Thread	74
Shared Memory/Block	0 B
Shared Memory Requested	64 KiB
Shared Memory Executed	64 KiB
Shared Memory Bank Size	4 B

[0] GeForce GTX 860M	
Compute Capability	5.0
Max. Threads per Block	1024
Max. Shared Memory per Block	48 KiB
Max. Registers per Block	65536
Max. Grid Dimensions	[2147483647, 65535, 65535]
Max. Block Dimensions	[1024, 1024, 64]
Max. Warps per Multiprocessor	64
Max. Blocks per Multiprocessor	32
Number of Multiprocessors	5
Multiprocessor Clock Rate	1.02 GHz
Concurrent Kernel	true
Max IPC	2
Threads per Warp	32
Global Memory Bandwidth	80.16 GB/s
Global Memory Size	4 GiB
Constant Memory Size	64 KiB
L2 Cache Size	2 MiB
Memcpy Engines	1
PCIe Generation	3
PCIe Link Rate	8 Gbit/s
PCIe Link Width	16

1. Compute, Bandwidth, or Latency Bound

The first step in analyzing an individual kernel is to determine if the performance of the kernel is bounded by computation, memo bandwidth, or instruction/memory latency. The results below indicate that the performance of kernel "atmosphereGPU" is most likely limited by compute. You should first examine the information in the "Compute Resources" section to determine how it is limiting performance.

1.1. Kernel Performance Is Bound By Compute

For device "GeForce GTX 860M" the kernel's memory utilization is significantly lower than its compute utilization. These utilization levels indicate that the performance of the kernel is most likely being limited by computation on the SMs.

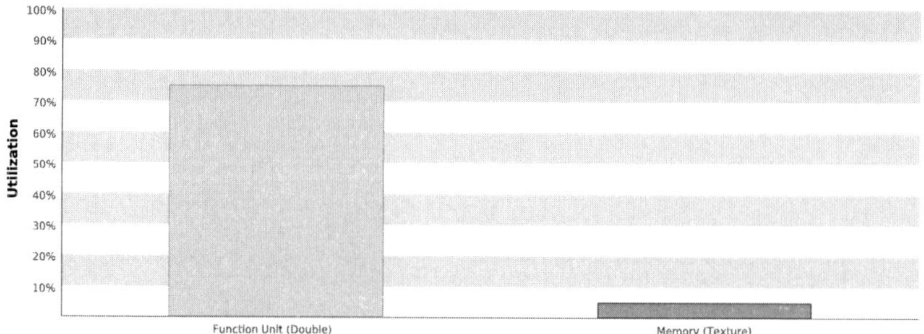

Compute Resources

U compute resources limit the performance of a kernel when those resources are insufficient or poorly utilized. Compute ources are used most efficiently when all threads in a warp have the same branching and predication behavior. The results below icate that a significant fraction of the available compute performance is being wasted because branch and predication behavior is 'ering for threads within a warp. Compute resources are used most efficiently when instructions do not overuse a function unit. e results below indicate that compute performance may be limited by overuse of a function unit.

. Low Warp Execution Efficiency

urp execution efficiency is the average percentage of active threads in each executed warp. Increasing warp execution efficiency l increase utilization of the GPU's compute resources. The kernel's warp execution efficiency of 18.6% is less than 100% due to ergent branches and predicated instructions. If predicated instructions are not taken into account the warp execution efficiency these kernels is 19.5%.

timization: Reduce the amount of intra-warp divergence and predication in the kernel.

. Divergent Branches

mpute resource are used most efficiently when all threads in a warp have the same branching behavior. When this does not cur the branch is said to be divergent. Divergent branches lower warp execution efficiency which leads to inefficient use of the U's compute resources.

timization: Each entry below points to a divergent branch within the kernel. For each branch reduce the amount of intra-warp ergence.

/home/marek/projects/repos/ikebana/src/ikebana/AtmosphericData.cu

ne 41	Divergence = 99.7% [1559 divergent executions out of 1563 total executions]
ne 44	Divergence = 30.7% [478 divergent executions out of 1559 total executions]

/home/marek/projects/repos/ikebana/src/ikebana/PerturbationAtmosphere.cu

ne 107	Divergence = 26.9% [420 divergent executions out of 1563 total executions]
ne 114	Divergence = 98.1% [1534 divergent executions out of 1563 total executions]
ne 134	Divergence = 95% [1485 divergent executions out of 1563 total executions]
ne 245	Divergence = 2.1% [1028 divergent executions out of 48839 total executions]
ne 292	Divergence = 4.8% [71 divergent executions out of 1485 total executions]
ne 293	Divergence = 5.2% [77 divergent executions out of 1481 total executions]
ne 294	Divergence = 5.5% [82 divergent executions out of 1478 total executions]
ne 295	Divergence = 4.3% [63 divergent executions out of 1471 total executions]

/usr/local/cuda-6.5/targets/x86_64-linux/include/math_functions.h

ne 9549	Divergence = 24.5% [665 divergent executions out of 2709 total executions]
ne 9549	Divergence = 12.4% [207 divergent executions out of 1667 total executions]
ne 9549	Divergence = 17.5% [207 divergent executions out of 1180 total executions]

. GPU Utilization Is Limited By Function Unit Usage

fferent types of instructions are executed on different function units within each SM. Performance can be limited if a function it is over-used by the instructions executed by the kernel. The following results show that the kernel's performance is potentially nited by overuse of the following function units: Double.

oad/Store - Load and store instructions for shared and constant memory.

exture - Load and store instructions for local, global, and texture memory.

Single - Single-precision integer and floating-point arithmetic instructions.
Double - Double-precision floating-point arithmetic instructions.
Special - Special arithmetic instructions such as sin, cos, popc, etc.
Control-Flow - Direct and indirect branches, jumps, and calls.

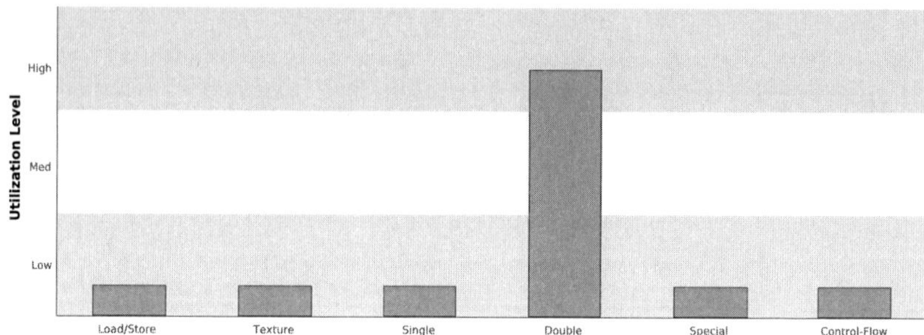

2.4. Instruction Execution Counts

The following chart shows the mix of instructions executed by the kernel. The instructions are grouped into classes and for each class the chart shows the percentage of thread execution cycles that were devoted to executing instructions in that class. The "Inactive" result shows the thread executions that did not execute any instruction because the thread was predicated or inactive du to divergence.

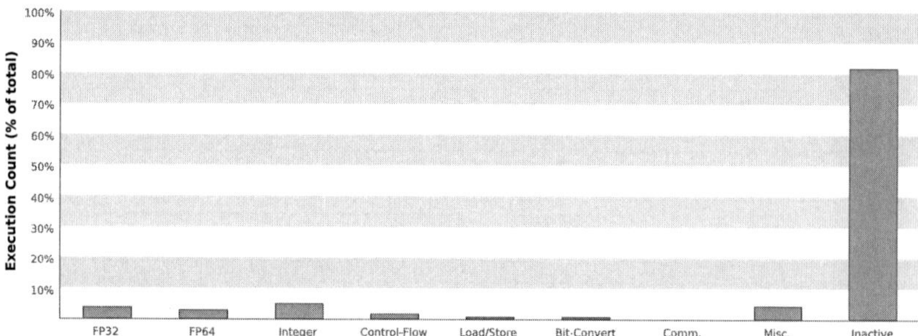

2.5. Floating-Point Operation Counts

The following chart shows the mix of floating-point operations executed by the kernel. The operations are grouped into classes an for each class the chart shows the percentage of thread execution cycles that were devoted to executing operations in that class. Th results do not sum to 100% because non-floating-point operations executed by the kernel are not shown in this chart.

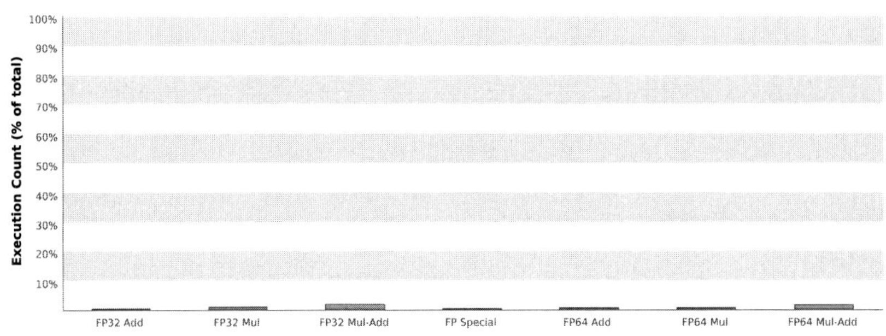
Execution Count (% of total)
100%
90%
80%
70%
60%
50%
40%
30%
20%
10%
FP32 Add
FP32 Mul
FP32 Mul-Add
FP Special
FP64 Add
FP64 Mul
FP64 Mul-Add

3. Memory Bandwidth

Memory bandwidth limits the performance of a kernel when one or more memories in the GPU cannot provide data at the rate requested by the kernel.

3.1. Memory Bandwidth And Utilization

The following table shows the memory bandwidth used by this kernel for the various types of memory on the device. The table al. shows the utilization of each memory type relative to the maximum throughput supported by the memory.

	Transactions	Bandwidth	Utilization
Shared Memory			
Shared Loads	0	0 B/s	
Shared Stores	0	0 B/s	
Shared Total	0	0 B/s	Idle Low Medium High Max
L2 Cache			
Reads	2543299	11.797 GB/s	
Writes	540865	2.509 GB/s	
Total	3084164	14.305 GB/s	Idle Low Medium High Max
Unified Cache			
Local Loads	92837	430.611 MB/s	
Local Stores	296769	1.377 GB/s	
Global Loads	2556472	11.858 GB/s	
Global Stores	244090	1.132 GB/s	
Texture Reads	557018	2.584 GB/s	
Unified Total	3747186	17.381 GB/s	Idle Low Medium High Max
Device Memory			
Reads	154891	718.44 MB/s	
Writes	238043	1.104 GB/s	
Total	392934	1.823 GB/s	Idle Low Medium High Max
System Memory			
[PCIe configuration: Gen3 x16, 8 Gbit/s]			
Reads	0	0 B/s	Idle Low Medium High Max
Writes	5	23.191 kB/s	Idle Low Medium High Max

Instruction and Memory Latency

truction and memory latency limit the performance of a kernel when the GPU does not have enough work to keep busy. The formance of latency-limited kernels can often be improved by increasing occupancy. Occupancy is a measure of how many ps the kernel has active on the GPU, relative to the maximum number of warps supported by the GPU. Theoretical occupancy vides an upper bound while achieved occupancy indicates the kernel's actual occupancy. The results below indicate that upancy can be improved by reducing the number of registers used by the kernel.

. GPU Utilization Is Limited By Register Usage

e kernel uses 74 registers for each thread (4736 registers for each block). This register usage is likely preventing the kernel from ly utilizing the GPU. Device "GeForce GTX 860M" provides up to 65536 registers for each block. Because the kernel uses 4736 isters for each block each SM is limited to simultaneously executing 12 blocks (24 warps). Chart "Varying Register Count" ow shows how changing register usage will change the number of blocks that can execute on each SM.

timization: Use the -maxrregcount flag or the __launch_bounds__ qualifier to decrease the number of registers used by each ead. This will increase the number of blocks that can execute on each SM.

Variable	Achieved	Theoretical	Device Limit	Grid Size: [782,1,1] (782 blocks) Block Size: [64,1,1]
Occupancy Per SM				
Active Blocks		12	32	0 4 8 12 16 20 24 28 32
Active Warps	22	24	64	0 9 18 27 36 45 54 6364
Active Threads		768	2048	0 512 1024 1536 2048
Occupancy	34.4%	37.5%	100%	0% 25% 50% 75% 100%
Warps				
Threads/Block		64	1024	0 256 512 768 1024
Warps/Block		2	32	0 4 8 12 16 20 24 28 32
Block Limit		32	32	0 4 8 12 16 20 24 28 32
Registers				
Registers/Thread		74	255	0 64 128 192 255
Registers/Block		5120	65536	0 16k 32k 48k 64k
Block Limit		12	32	0 4 8 12 16 20 24 28 32
Shared Memory				
Shared Memory/Block		0	65536	0 16k 32k 48k 64k
Block Limit			32	

. Occupancy Charts

e following charts show how varying different components of the kernel will impact theoretical occupancy.

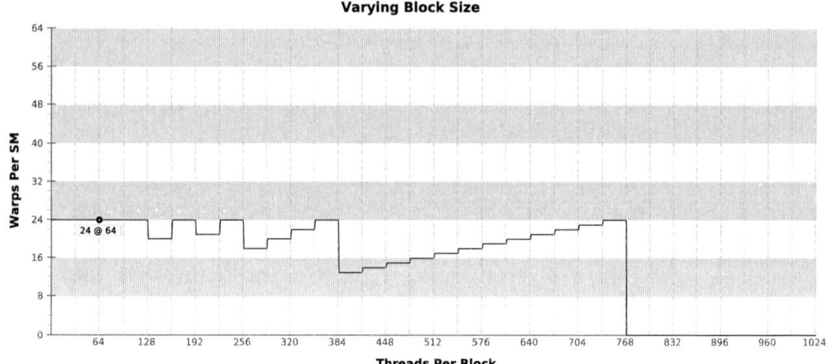
Varying Block Size
Warps Per SM
Threads Per Block
24 @ 64
64
56
48
40
32
24
16
8
0
64 128 192 256 320 384 448 512 576 640 704 768 832 896 960 1024

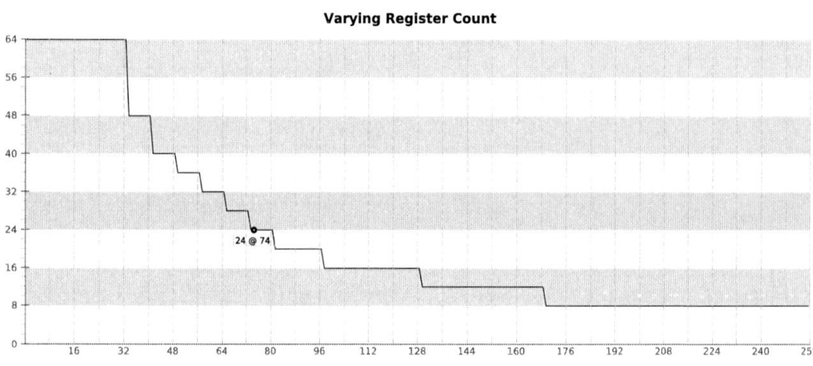
Varying Register Count
Registers Per Thread
24 @ 74
64
56
48
40
32
24
16
8
0
16 32 48 64 80 96 112 128 144 160 176 192 208 224 240 255

Varying Shared Memory Usage

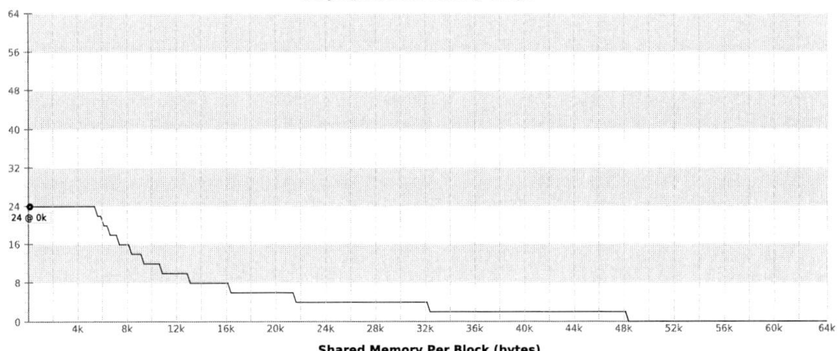
64
56
48
40
32
24
24 @ 0k
16
8
0
4k
8k
12k
16k
20k
24k
28k
32k
36k
40k
44k
48k
52k
56k
60k
64k
Shared Memory Per Block (bytes)

A.2. GeForce GTX 960

Analysis Report

atmosphereGPU(OPI::Orbit*, OPI::ObjectProperties*, OPI::Orbit*, int, tAtmoData*, int*, double*, float, float)

Duration	3.299 ms (3,298,552 ns)
Grid Size	[782,1,1]
Block Size	[64,1,1]
Registers/Thread	62
Shared Memory/Block	0 B
Shared Memory Requested	96 KiB
Shared Memory Executed	96 KiB
Shared Memory Bank Size	4 B

[0] GeForce GTX 960

GPU UUID	GPU-206126c3-782c-6757-35bb-74f09da6b657
Compute Capability	5.2
Max. Threads per Block	1024
Max. Shared Memory per Block	48 KiB
Max. Registers per Block	65536
Max. Grid Dimensions	[2147483647, 65535, 65535]
Max. Block Dimensions	[1024, 1024, 64]
Max. Warps per Multiprocessor	64
Max. Blocks per Multiprocessor	32
Single Precision FLOP/s	2.618 TeraFLOP/s
Double Precision FLOP/s	0 FLOP/s
Number of Multiprocessors	8
Multiprocessor Clock Rate	1.278 GHz
Concurrent Kernel	true
Max IPC	6
Threads per Warp	32
Global Memory Bandwidth	112.16 GB/s
Global Memory Size	3.999 GiB
Constant Memory Size	64 KiB
L2 Cache Size	1 MiB
Memcpy Engines	2
PCIe Generation	2
PCIe Link Rate	5 Gbit/s
PCIe Link Width	16

Compute, Bandwidth, or Latency Bound

e first step in analyzing an individual kernel is to determine if the performance of the kernel is bounded by computation, memory dwidth, or instruction/memory latency. The results below indicate that the performance of kernel "atmosphereGPU" is most ely limited by compute. You should first examine the information in the "Compute Resources" section to determine how it is iting performance.

. Kernel Performance Is Bound By Compute

r device "GeForce GTX 960" the kernel's memory utilization is significantly lower than its compute utilization. These utilization els indicate that the performance of the kernel is most likely being limited by computation on the SMs.

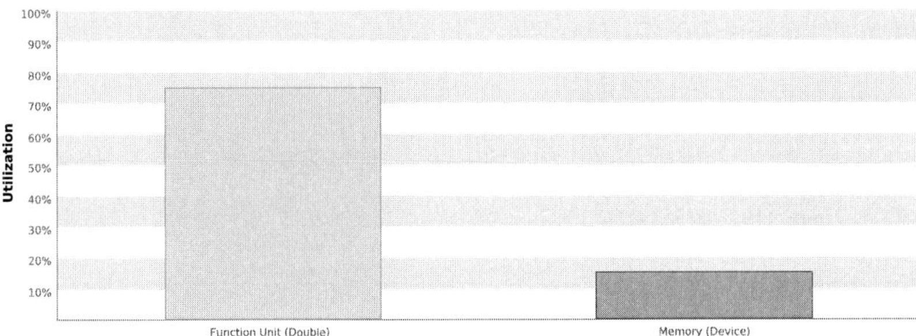

2. Compute Resources

GPU compute resources limit the performance of a kernel when those resources are insufficient or poorly utilized. Compute resources are used most efficiently when all threads in a warp have the same branching and predication behavior. The results below indicate that a significant fraction of the available compute performance is being wasted because branch and predication behavior differing for threads within a warp. Compute resources are used most efficiently when instructions do not overuse a function unit. The results below indicate that compute performance may be limited by overuse of a function unit.

2.1. Low Warp Execution Efficiency

Warp execution efficiency is the average percentage of active threads in each executed warp. Increasing warp execution efficiency will increase utilization of the GPU's compute resources. The kernel's warp execution efficiency of 19.4% is less than 100% due to divergent branches and predicated instructions. If predicated instructions are not taken into account the warp execution efficiency for these kernels is 21.1%.

Optimization: Reduce the amount of intra-warp divergence and predication in the kernel.

2.2. Divergent Branches

Compute resource are used most efficiently when all threads in a warp have the same branching behavior. When this does not occur the branch is said to be divergent. Divergent branches lower warp execution efficiency which leads to inefficient use of the GPU's compute resources.

Optimization: Each entry below points to a divergent branch within the kernel. For each branch reduce the amount of intra-warp divergence.

/home/wansti/work/ikebana/src/ikebana/AtmosphericData.cu	
Line 41	Divergence = 99.7% [1559 divergent executions out of 1563 total executions]
Line 44	Divergence = 31.2% [486 divergent executions out of 1559 total executions]

/home/wansti/work/ikebana/src/ikebana/PerturbationAtmosphere.cu	
Line 107	Divergence = 10.2% [160 divergent executions out of 1563 total executions]
Line 114	Divergence = 98.1% [1534 divergent executions out of 1563 total executions]
Line 134	Divergence = 95.1% [1487 divergent executions out of 1563 total executions]
Line 245	Divergence = 2.1% [1042 divergent executions out of 48910 total executions]

/usr/local/cuda/bin/../targets/x86_64-linux/include/math_functions.hpp	
Line 1324	Divergence = 14.4% [168 divergent executions out of 1166 total executions]
Line 1324	Divergence = 10.5% [168 divergent executions out of 1607 total executions]
Line 1324	Divergence = 27.4% [744 divergent executions out of 2718 total executions]

2.3. GPU Utilization Is Limited By Function Unit Usage

Different types of instructions are executed on different function units within each SM. Performance can be limited if a function unit is over-used by the instructions executed by the kernel. The following results show that the kernel's performance is potentially limited by overuse of the following function units: Double.

Load/Store - Load and store instructions for shared and constant memory.
Texture - Load and store instructions for local, global, and texture memory.
Single - Single-precision integer and floating-point arithmetic instructions.
Double - Double-precision floating-point arithmetic instructions.
Special - Special arithmetic instructions such as sin, cos, popc, etc.
Control-Flow - Direct and indirect branches, jumps, and calls.

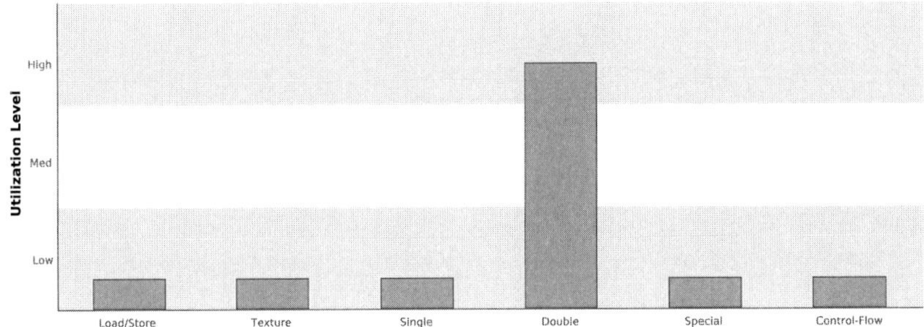

. Instruction Execution Counts

e following chart shows the mix of instructions executed by the kernel. The instructions are grouped into classes and for each ss the chart shows the percentage of thread execution cycles that were devoted to executing instructions in that class. The active" result shows the thread executions that did not execute any instruction because the thread was predicated or inactive due divergence.

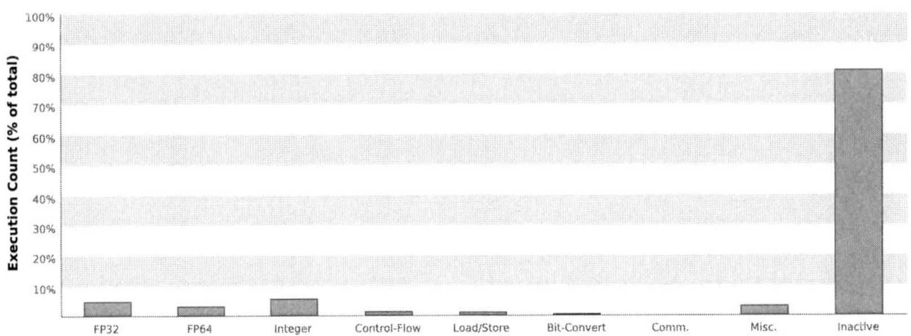

. Floating-Point Operation Counts

e following chart shows the mix of floating-point operations executed by the kernel. The operations are grouped into classes and each class the chart shows the percentage of thread execution cycles that were devoted to executing operations in that class. The ults do not sum to 100% because non-floating-point operations executed by the kernel are not shown in this chart.

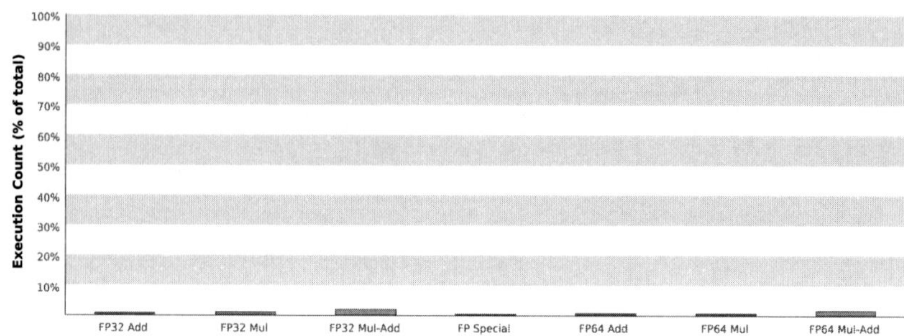
Execution Count (% of total)
100%
90%
80%
70%
60%
50%
40%
30%
20%
10%
FP32 Add
FP32 Mul
FP32 Mul-Add
FP Special
FP64 Add
FP64 Mul
FP64 Mul-Add

Memory Bandwidth

mory bandwidth limits the performance of a kernel when one or more memories in the GPU cannot provide data at the rate uested by the kernel.

. Memory Bandwidth And Utilization

e following table shows the memory bandwidth used by this kernel for the various types of memory on the device. The table also ws the utilization of each memory type relative to the maximum throughput supported by the memory.

	Transactions	Bandwidth	Utilization
Shared Memory			
Shared Loads	0	0 B/s	
Shared Stores	0	0 B/s	
Shared Total	0	0 B/s	Idle Low Medium High Max
L2 Cache			
Reads	2562943	23.632 GB/s	
Writes	586513	5.408 GB/s	
Total	3149456	29.04 GB/s	Idle Low Medium High Max
Unified Cache			
Local Loads	162764	1.501 GB/s	
Local Stores	340876	3.143 GB/s	
Global Loads	2581989	23.808 GB/s	
Global Stores	245630	2.265 GB/s	
Texture Reads	599954	5.532 GB/s	
Unified Total	3931213	36.248 GB/s	Idle Low Medium High Max
Device Memory			
Reads	749349	6.91 GB/s	
Writes	700971	6.463 GB/s	
Total	1450320	13.373 GB/s	Idle Low Medium High Max
System Memory			
[PCIe configuration: Gen2 x16, 5 Gbit/s]			
Reads	28	258.179 kB/s	Idle Low Medium High Max
Writes	5	46.103 kB/s	Idle Low Medium High Max

4. Instruction and Memory Latency

Instruction and memory latency limit the performance of a kernel when the GPU does not have enough work to keep busy. The performance of latency-limited kernels can often be improved by increasing occupancy. Occupancy is a measure of how many warps the kernel has active on the GPU, relative to the maximum number of warps supported by the GPU. Theoretical occupancy provides an upper bound while achieved occupancy indicates the kernel's actual occupancy. The results below indicate that occupancy can be improved by reducing the number of registers used by the kernel.

4.1. GPU Utilization May Be Limited By Register Usage

Theoretical occupancy is less than 100% but is large enough that increasing occupancy may not improve performance. You can attempt the following optimization to increase the number of warps on each SM but it may not lead to increased performance.

The kernel uses 62 registers for each thread (3968 registers for each block). This register usage is likely preventing the kernel fror fully utilizing the GPU. Device "GeForce GTX 960" provides up to 65536 registers for each block. Because the kernel uses 3968 registers for each block each SM is limited to simultaneously executing 16 blocks (32 warps). Chart "Varying Register Count" below shows how changing register usage will change the number of blocks that can execute on each SM.

Optimization: Use the -maxrregcount flag or the __launch_bounds__ qualifier to decrease the number of registers used by each thread. This will increase the number of blocks that can execute on each SM. On devices with Compute Capability 5.2 turning global cache off can increase the occupancy limited by register usage.

Variable	Achieved	Theoretical	Device Limit	Grid Size: [782,1,1] (782 blocks) Block Size: [64,1,1] (64 threads
Occupancy Per SM				
Active Blocks		16	32	0 3 6 9 12 15 18 21 24 27 30 32
Active Warps	29.52	32	64	0 7 14 21 28 35 42 49 56 6684
Active Threads		1024	2048	0 256 512 768 1024 1280 1536 1792 2048
Occupancy	46.1%	50%	100%	0% 25% 50% 75% 100%
Warps				
Threads/Block		64	1024	0 128 256 384 512 640 768 896 1024
Warps/Block		2	32	0 3 6 9 12 15 18 21 24 27 30 32
Block Limit		32	32	0 3 6 9 12 15 18 21 24 27 30 32
Registers				
Registers/Thread		62	255	0 32 64 96 128 160 192 224 255
Registers/Block		4096	65536	0 16k 32k 48k 64k
Block Limit		16	32	0 3 6 9 12 15 18 21 24 27 30 32
Shared Memory				
Shared Memory/Block		0	98304	0 32k 64k 96k
Block Limit			32	

4.2. Occupancy Charts

The following charts show how varying different components of the kernel will impact theoretical occupancy.

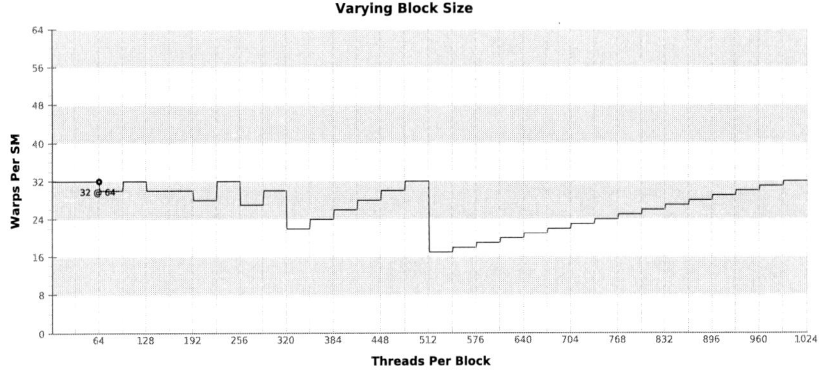
Varying Block Size
Warps Per SM
64
56
48
40
32
24
16
8
0
32 @ 64
64
128
192
256
320
384
448
512
576
640
704
768
832
896
960
1024
Threads Per Block

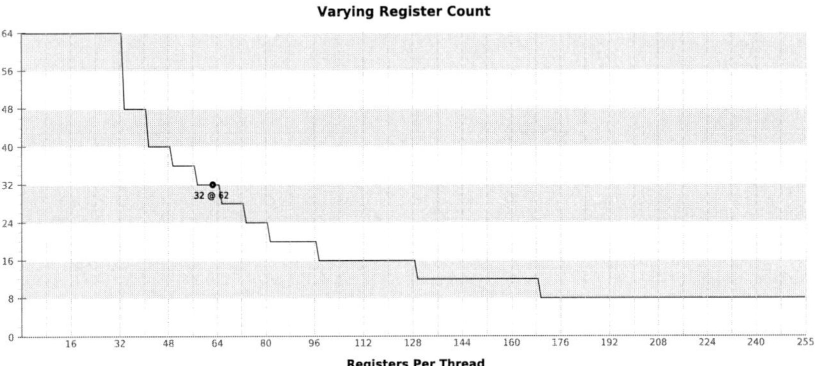
Varying Register Count
64
56
48
40
32
24
16
8
0
32 @ 62
16
32
48
64
80
96
112
128
144
160
176
192
208
224
240
255
Registers Per Thread

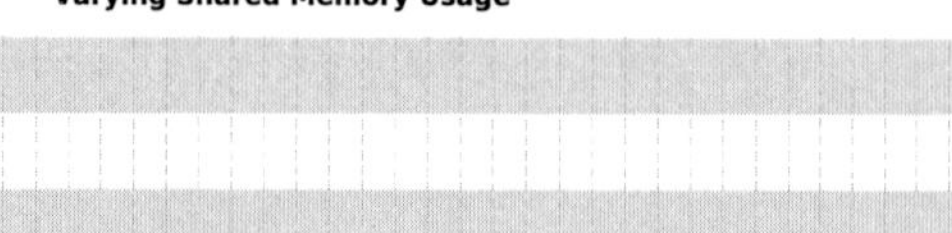
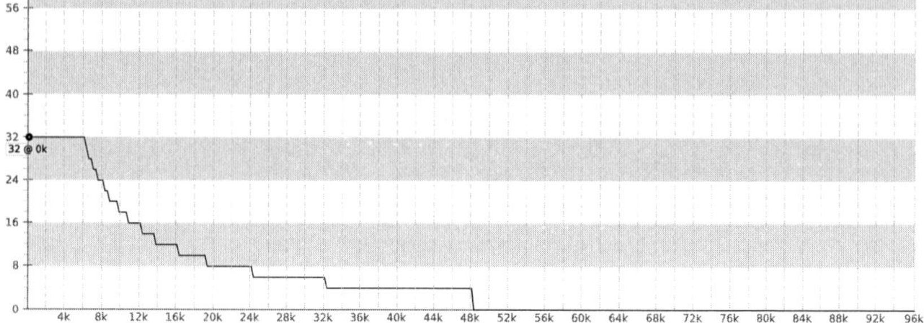
Varying Shared Memory Usage
64
56
48
40
32
32 @ 0k
24
16
8
0
4k 8k 12k 16k 20k 24k 28k 32k 36k 40k 44k 48k 52k 56k 60k 64k 68k 72k 76k 80k 84k 88k 92k 96k
Shared Memory Per Block (bytes)

A.3. Tesla K20c

Analysis Report

atmosphereGPU(OPI::Orbit*, OPI::ObjectProperties*, OPI::Orbit*, int, tAtmoData*, int*, double*, float, float)

Duration	17.818 ms (17,817,977 ns)
Grid Size	[782,1,1]
Block Size	[64,1,1]
Registers/Thread	76
Shared Memory/Block	0 B
Shared Memory Requested	48 KiB
Shared Memory Executed	48 KiB
Shared Memory Bank Size	4 B

[0] Tesla K20c

Compute Capability	3.5
Max. Threads per Block	1024
Max. Shared Memory per Block	48 KiB
Max. Registers per Block	65536
Max. Grid Dimensions	[2147483647, 65535, 65535]
Max. Block Dimensions	[1024, 1024, 64]
Max. Warps per Multiprocessor	64
Max. Blocks per Multiprocessor	16
Number of Multiprocessors	13
Multiprocessor Clock Rate	705.5 MHz
Concurrent Kernel	true
Max IPC	7
Threads per Warp	32
Global Memory Bandwidth	208 GB/s
Global Memory Size	4.687 GiB
Constant Memory Size	64 KiB
L2 Cache Size	1.25 MiB
Memcpy Engines	2
PCIe Generation	2
PCIe Link Rate	5 Gbit/s
PCIe Link Width	16

1. Compute, Bandwidth, or Latency Bound

The first step in analyzing an individual kernel is to determine if the performance of the kernel is bounded by computation, memo bandwidth, or instruction/memory latency. The results below indicate that the performance of kernel "atmosphereGPU" is most likely limited by instruction and memory latency. You should first examine the information in the "Instruction And Memory Latency" section to determine how it is limiting performance.

1.1. Kernel Performance Is Bound By Instruction And Memory Latency

This kernel exhibits low compute throughput and memory bandwidth utilization relative to the peak performance of "Tesla K20c' These utilization levels indicate that the performance of the kernel is most likely limited by the latency of arithmetic or memory operations. Achieved compute throughput and/or memory bandwidth below 60% of peak typically indicates latency issues.

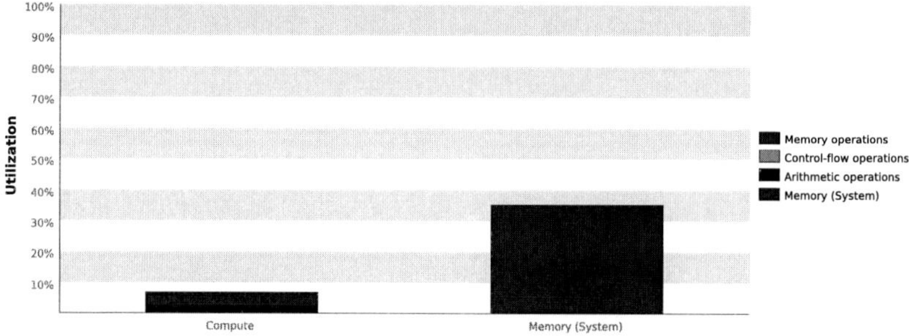

Instruction and Memory Latency

ruction and memory latency limit the performance of a kernel when the GPU does not have enough work to keep busy. The formance of latency-limited kernels can often be improved by increasing occupancy. Occupancy is a measure of how many ps the kernel has active on the GPU, relative to the maximum number of warps supported by the GPU. Theoretical occupancy vides an upper bound while achieved occupancy indicates the kernel's actual occupancy. The results below indicate that upancy can be improved by reducing the number of registers used by the kernel.

. GPU Utilization Is Limited By Register Usage

e kernel uses 76 registers for each thread (4864 registers for each block). This register usage is likely preventing the kernel from ly utilizing the GPU. Device "Tesla K20c" provides up to 65536 registers for each block. Because the kernel uses 4864 registers each block each SM is limited to simultaneously executing 12 blocks (24 warps). Chart "Varying Register Count" below shows v changing register usage will change the number of blocks that can execute on each SM.

timization: Use the -maxrregcount flag or the __launch_bounds__ qualifier to decrease the number of registers used by each ead. This will increase the number of blocks that can execute on each SM.

Variable	Achieved	Theoretical	Device Limit	Grid Size: [782,1,1] (782 blocks) Block Size: [64,1,1]
Occupancy Per SM				
Active Blocks		12	16	0 2 4 6 8 10 12 14 16
				0 9 18 27 36 45 54 64
Active Threads		768	2048	0 512 1024 1536 2048
				0% 25% 50% 75% 100%
Warps				
Threads/Block		64	1024	0 256 512 768 1024
				0 4 8 12 16 20 24 28 32
Block Limit		32	16	0 2 4 6 8 10 12 14 16
Registers				
Registers/Thread		76	255	0 64 128 192 255
				0 16k 32k 48k 64k
Block Limit		12	16	0 2 4 6 8 10 12 14 16
Shared Memory				
Shared Memory/Block		0	49152	0 16k 32k 48k

2. Occupancy Charts

e following charts show how varying different components of the kernel will impact theoretical occupancy.

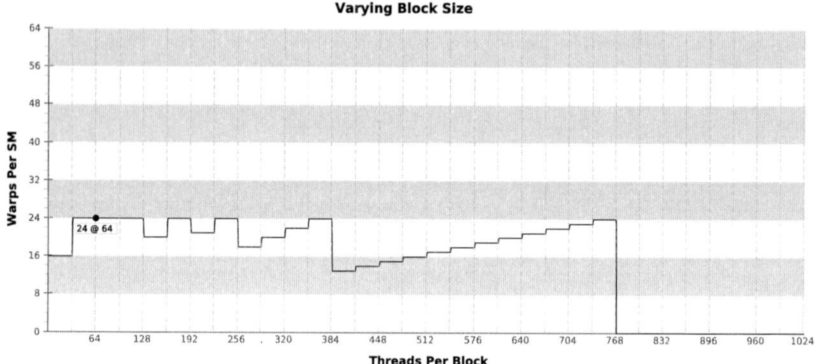
Varying Block Size
Warps Per SM
Threads Per Block
24 @ 64
0
8
16
24
32
40
48
56
64
128
192
256
320
384
448
512
576
640
704
768
832
896
960
1024

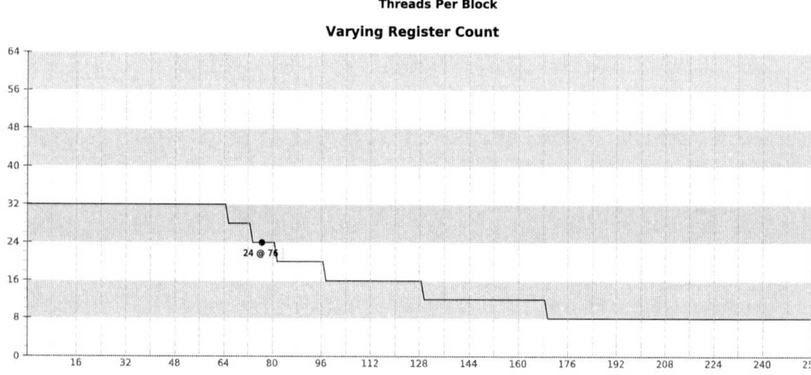
Varying Register Count
Registers Per Thread
24 @ 76
0
8
16
24
32
40
48
56
64
80
96
112
128
144
160
176
192
208
224
240
255

Varying Shared Memory Usage

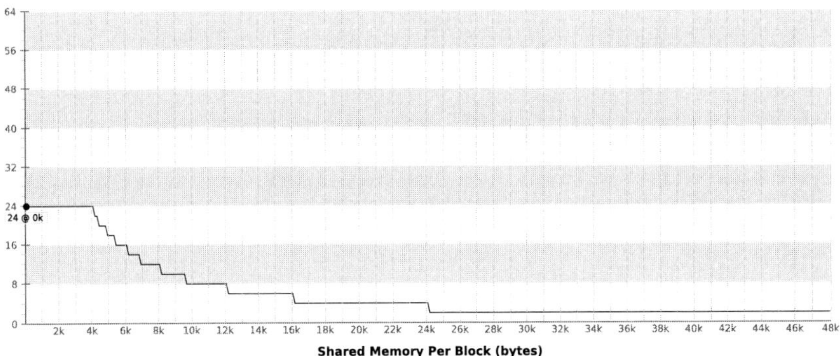
64
56
48
40
32
24
16
8
0
24 @ 0k
2k 4k 6k 8k 10k 12k 14k 16k 18k 20k 22k 24k 26k 28k 30k 32k 34k 36k 38k 40k 42k 44k 46k 48k
Shared Memory Per Block (bytes)

3. Compute Resources

GPU compute resources limit the performance of a kernel when those resources are insufficient or poorly utilized. Compute resources are used most efficiently when all threads in a warp have the same branching and predication behavior. The results belo indicate that a significant fraction of the available compute performance is being wasted because branch and predication behavior differing for threads within a warp.

3.1. Low Warp Execution Efficiency

Warp execution efficiency is the average percentage of active threads in each executed warp. Increasing warp execution efficienc will increase utilization of the GPU's compute resources. The kernel's warp execution efficiency of 19.4% is less than 100% due t divergent branches and predicated instructions. If predicated instructions are not taken into account the warp execution efficiency for these kernels is 20.5%.

Optimization: Reduce the amount of intra-warp divergence and predication in the kernel.

3.2. Divergent Branches

Compute resource are used most efficiently when all threads in a warp have the same branching behavior. When this does not occur the branch is said to be divergent. Divergent branches lower warp execution efficiency which leads to inefficient use of the GPU's compute resources.

Optimization: Each entry below points to a divergent branch within the kernel. For each branch reduce the amount of intra-warp divergence.

/home/users/mmoeckel/projects/repos/ikebana/src/ikebana/AtmosphericData.cu

Line 41	Divergence = 99.7% [1558 divergent executions out of 1563 total executions]
Line 44	Divergence = 30.6% [477 divergent executions out of 1558 total executions]

/home/users/mmoeckel/projects/repos/ikebana/src/ikebana/PerturbationAtmosphere.cu

Line 107	Divergence = 22.5% [351 divergent executions out of 1563 total executions]
Line 114	Divergence = 98.1% [1534 divergent executions out of 1563 total executions]
Line 134	Divergence = 95.1% [1486 divergent executions out of 1563 total executions]
Line 245	Divergence = 2.1% [1034 divergent executions out of 48874 total executions]
Line 292	Divergence = 4.7% [70 divergent executions out of 1486 total executions]
Line 293	Divergence = 4.5% [67 divergent executions out of 1484 total executions]
Line 294	Divergence = 6.1% [91 divergent executions out of 1480 total executions]
Line 295	Divergence = 4.4% [65 divergent executions out of 1472 total executions]

/usr/local/cuda/bin/..//include/math_functions.h

Line 9549	Divergence = 25.8% [694 divergent executions out of 2692 total executions]
Line 9549	Divergence = 16.8% [202 divergent executions out of 1204 total executions]
Line 9549	Divergence = 12% [202 divergent executions out of 1690 total executions]

3.3. Function Unit Utilization

Different types of instructions are executed on different function units within each SM. Performance can be limited if a function unit is over-used by the instructions executed by the kernel. The following results show that the kernel's performance is not limite by overuse of any function unit.

Load/Store - Load and store instructions for local, shared, global, constant, etc. memory.
Arithmetic - All arithmetic instructions including integer and floating-point add and multiply, logical and binary operations, etc.
Control-Flow - Direct and indirect branches, jumps, and calls.

xture - Texture operations.

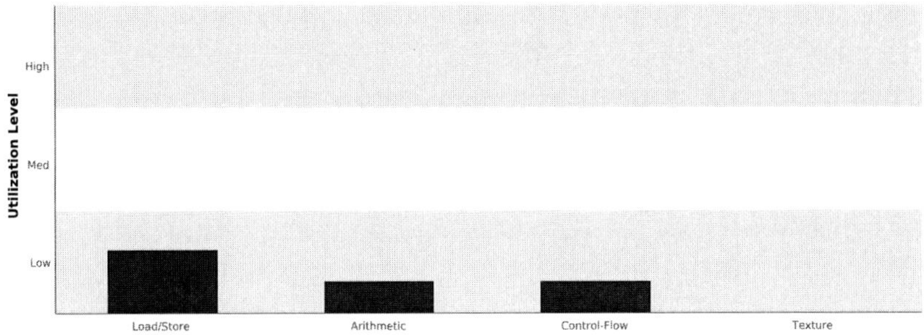

. Instruction Execution Counts

e following chart shows the mix of instructions executed by the kernel. The instructions are grouped into classes and for each ss the chart shows the percentage of thread execution cycles that were devoted to executing instructions in that class. The active" result shows the thread executions that did not execute any instruction because the thread was predicated or inactive due divergence.

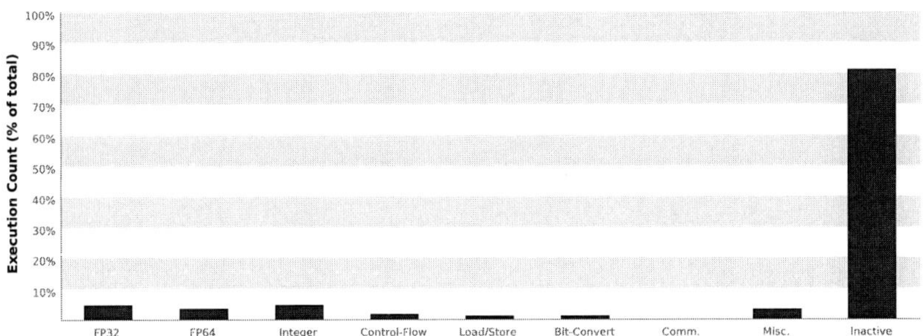

. Floating-Point Operation Counts

e following chart shows the mix of floating-point operations executed by the kernel. The operations are grouped into classes and each class the chart shows the percentage of thread execution cycles that were devoted to executing operations in that class. The ults do not sum to 100% because non-floating-point operations executed by the kernel are not shown in this chart.

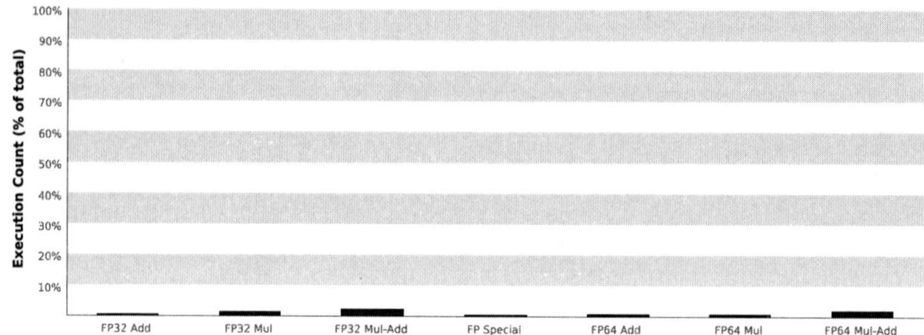
100%
90%
80%
70%
60%
50%
40%
30%
20%
10%
Execution Count (% of total)
FP32 Add
FP32 Mul
FP32 Mul-Add
FP Special
FP64 Add
FP64 Mul
FP64 Mul-Add

Memory Bandwidth

mory bandwidth limits the performance of a kernel when one or more memories in the GPU cannot provide data at the rate uested by the kernel.

. Memory Bandwidth And Utilization

e following table shows the memory bandwidth used by this kernel for the various types of memory on the device. The table also ows the utilization of each memory type relative to the maximum throughput supported by the memory.

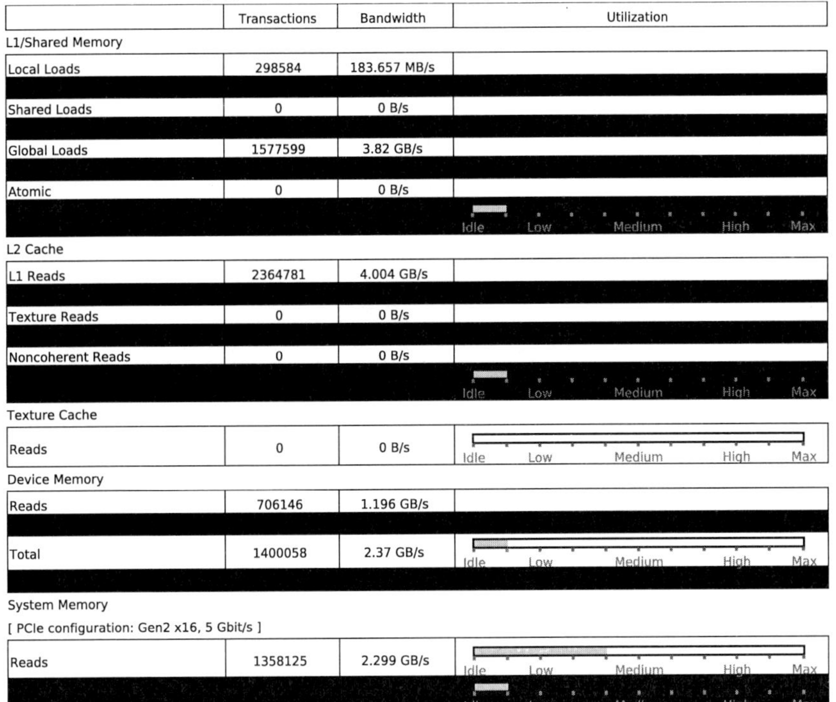

	Transactions	Bandwidth	Utilization
L1/Shared Memory			
Local Loads	298584	183.657 MB/s	
Shared Loads	0	0 B/s	
Global Loads	1577599	3.82 GB/s	
Atomic	0	0 B/s	Idle Low Medium High Max
L2 Cache			
L1 Reads	2364781	4.004 GB/s	
Texture Reads	0	0 B/s	
Noncoherent Reads	0	0 B/s	Idle Low Medium High Max
Texture Cache			
Reads	0	0 B/s	Idle Low Medium High Max
Device Memory			
Reads	706146	1.196 GB/s	
Total	1400058	2.37 GB/s	Idle Low Medium High Max
System Memory			
[PCIe configuration: Gen2 x16, 5 Gbit/s]			
Reads	1358125	2.299 GB/s	Idle Low Medium High Max Idle Low Medium High Max

B Individual Error Rate Plots

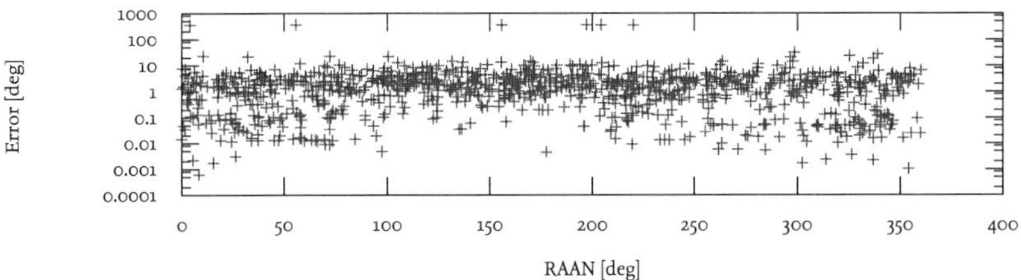

Figure B.1.: Error rates of the zonal harmonics module.

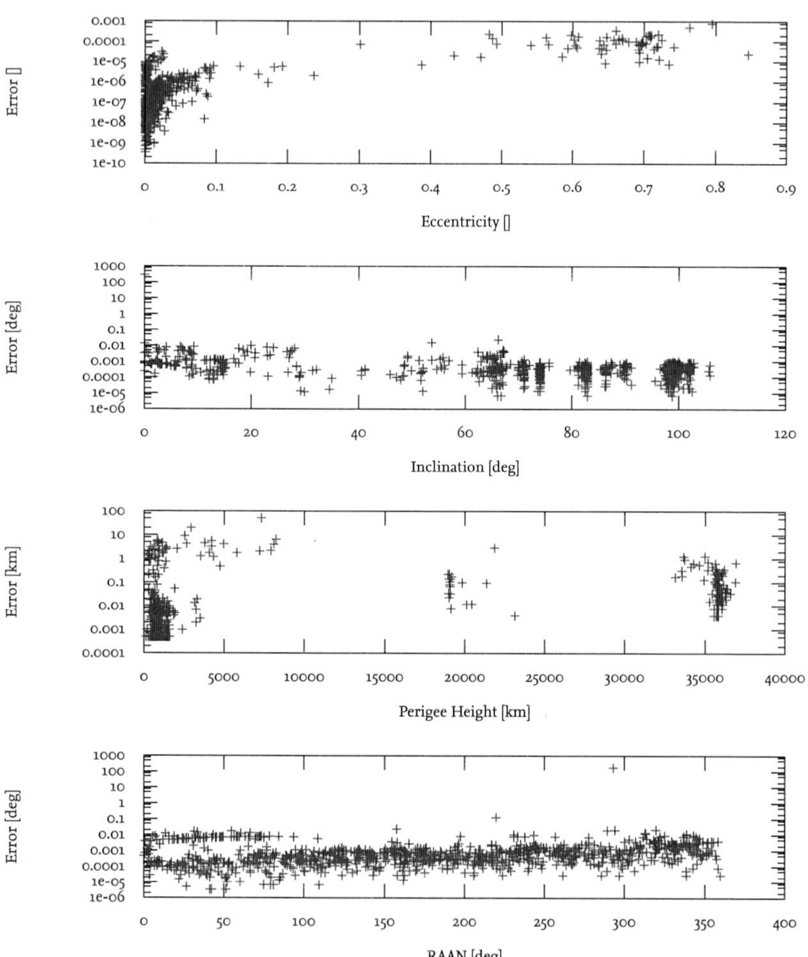

Figure B.2.: Error rates of the lunisolar module.

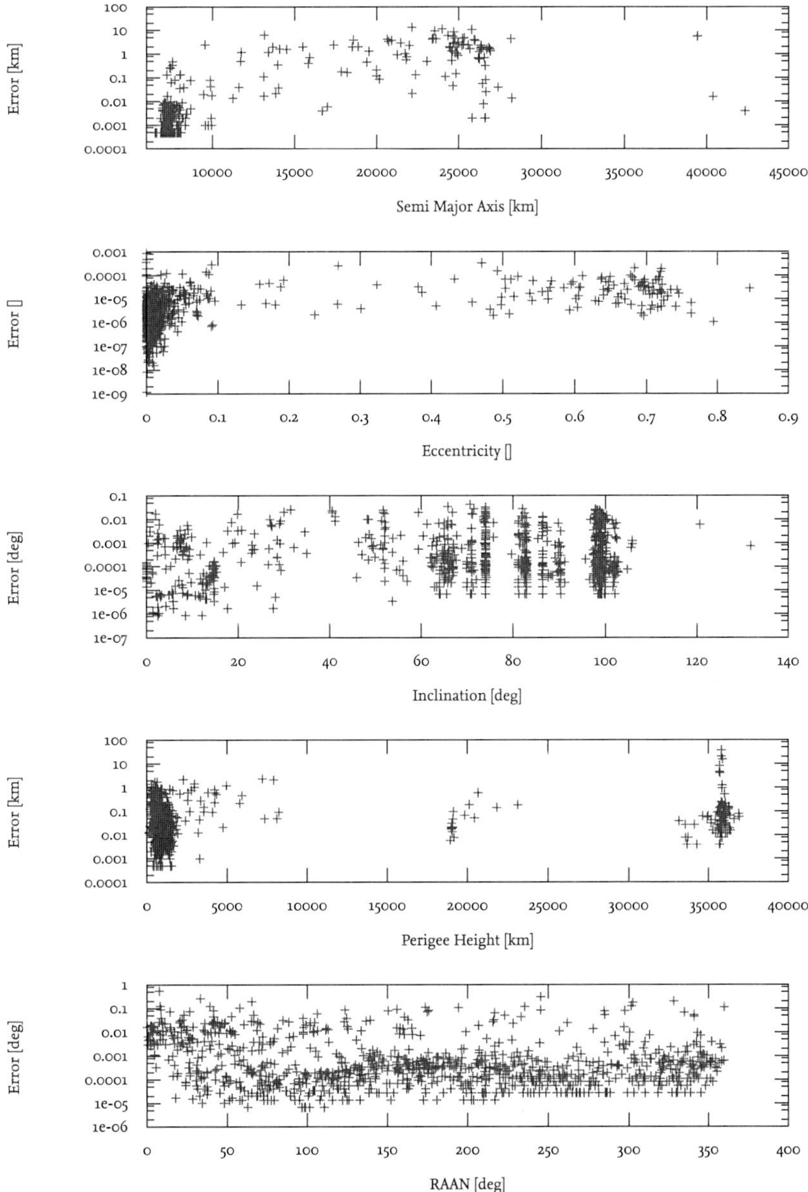

Figure B.3.: Error rates of the solar radiation pressure module.

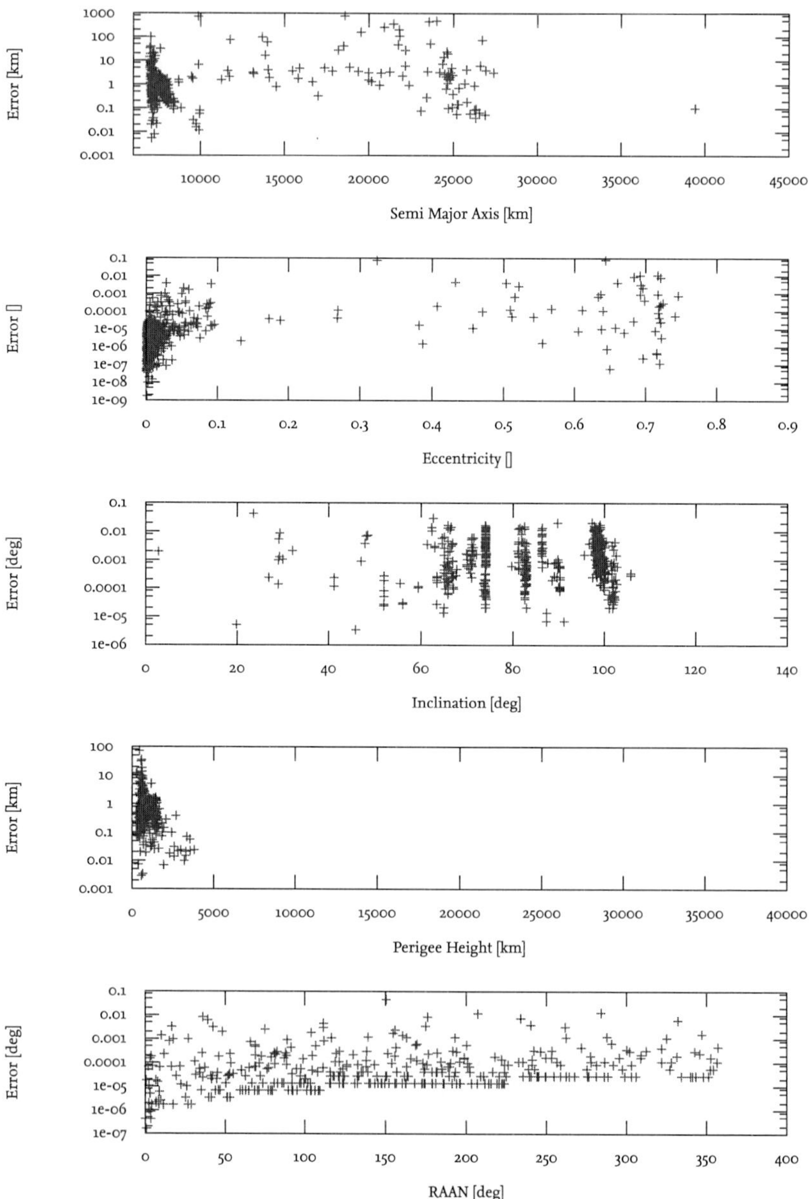

Figure B.4.: Error rates of the atmospherical module.

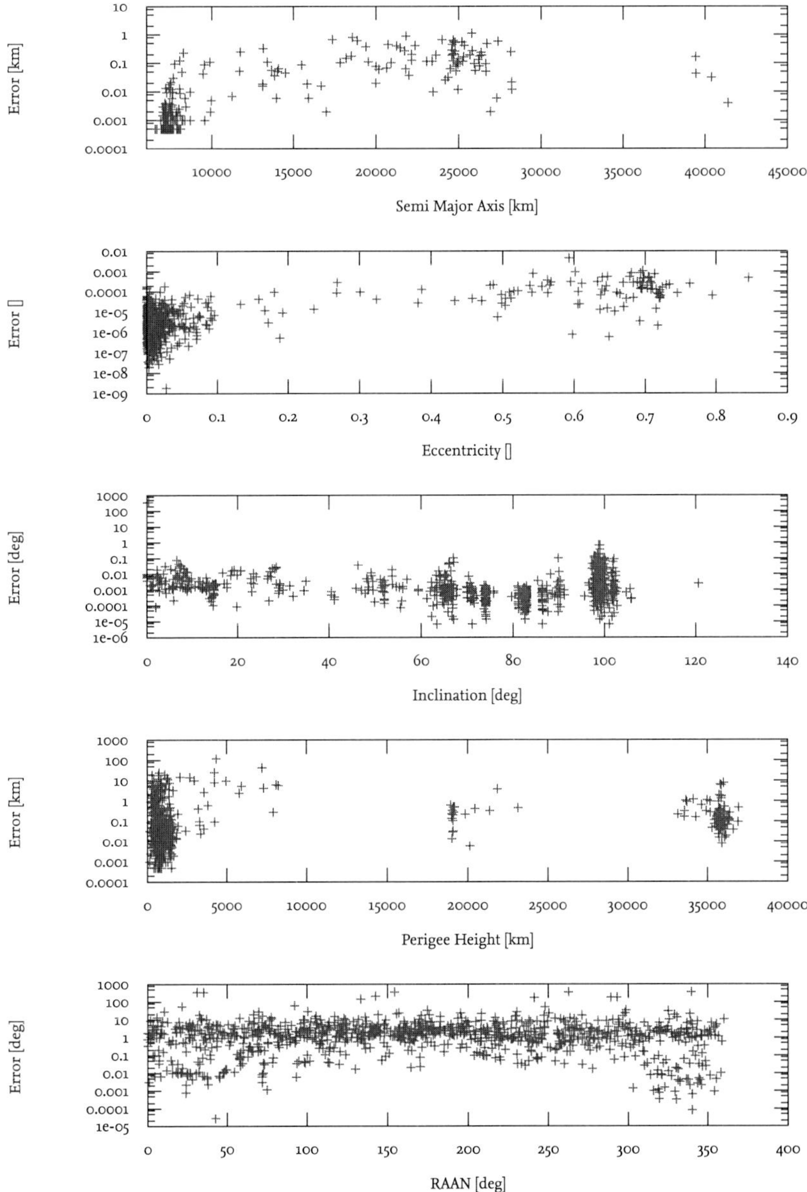

Figure B.5.: Combined error rates of zonal harmonics, lunisolar and solar radiation pressure modules.

C Ikebana Class Headers

C.1. Ikebana::Ikebana

```
#ifndef __IKEBANA_H__
#define __IKEBANA_H__

#define OPI_CUDA_PREFIX __host__ __device__

#include "OPI/opi_cpp.h"

// Plugin information
#define OPI_PLUGIN_NAME "Ikebana"
#define OPI_PLUGIN_AUTHOR "M. Moeckel, ILR, TU-BS"
#define OPI_PLUGIN_DESC "FLORA, rearranged"

// Plugin version
#define OPI_PLUGIN_VERSION_MAJOR 0
#define OPI_PLUGIN_VERSION_MINOR 1
#define OPI_PLUGIN_VERSION_PATCH 0

#include "PMMeanMotion.h"
#include "PMZonalHarmonics.h"
#include "PMLuniSolar.h"
#include "PMAtmosphere.h"
#include "PMSolarRadiation.h"

/*
                              #                      # #        #
                      #       #           #       #  # #        #
  #                   #       #           #       #             #      ##
  #        #         ##       #          ##       #          #######    ##
  #        ##        #   ##########      #   ##########         #        ##
  #         ##       #        #          #        #            ##
  #          #       #        #          #        #            #      #
  #          ##      #        #          #        #            #      #
  #           #      #        #          #        #           ##      #
  #           #      #        #          #        #           #       #
  ##  #       #      # #      #          #    #####          ##       #
   # ##              ##      ##          #   #    ###             #####
   ###                #      #           # # #    # ##           #    ###
    #                       ##           ##  #   ##  ##          #   ## ##
                           ##             #   ####                ####
-----------------------------------------------------------------------------
                      F L O R A  -  R E A R R A N G E D
*/


//! A CUDA-capable, analytical OPI Propagator.
/*! Ikebana is a port of the analytical propagator FLORA to C++ with the main
 * functionality being implemented as CUDA kernels. It uses the OPI interface
 * for easy integration into host applications that support it.
 */
class Ikebana: public OPI::Propagator
{
  public:
    //! Class contructor.
    /*!
```

```
     * This class implements the OPI::Propagator interface that enables it to be
     * compiled and used as an OPI plugin.
     * It is used to register OPI properties and set initial values for all variables.
     * @param host A pointer to the OPI::Host that calls this OPI::Propagator.
     */
    Ikebana(OPI::Host& host);

    //! Class destructor. Currently without function.
    ~Ikebana();

    //! OPI function that returns the minimum CUDA compute capability required to run.
    /*! @return 3, as this is the minimum CC required for Ikebana.
     */
    int requiresCUDA();

    //! OPI function that returns whether this propagator can propagate backwards.
    /*! @return false (until backward propagation is properly implemented and tested).
     */
    bool backwardPropagation();

protected:
    //! OPI-defined function to start propagation for a given Population and time.
    /*!
     * It takes as argument an instance of an OPI::Population containing data of orbital
     * objects that are to be propagated. The function starts propagation at the given
     * Julian Day and calculates the objects' absolute positions after the given amount
     * of seconds (dt) have passed. This function is ususally called multiple times by
     * the host in a loop over a specified time frame. The value of dt is usually the
     * amount of seconds that have passed between the Julian Days of two subsequent calls.
     * @param julian_day The date of the beginning of this propagation step.
     * @param dt The amount of seconds to be propagated.
     * @return OPI::SUCCESS or an error message.
     */
    OPI::ErrorCode runPropagation(OPI::Population& data, double julian_day, float dt);

    //! Callback for initialization.
    /*!
     * This function is called automatically when the host program selects and enables
     * this propagator. It currently has no function except printing a debug message but
     * it can be used for initialization steps that do not require any input data.
     * @return OPI::SUCCESS or an error message.
     */
    OPI::ErrorCode runEnable();

    //! Callback for deinitialization.
    /*!
     * Similar to runEnable(), this function is called automatically when the host
     * program disables this propagator. It is used to clean up allocated memory.
     * @return OPI::SUCCESS or an error message.
     */
    OPI::ErrorCode runDisable();

private:
    //! Variable to store the OPI Property "jTermLevel" for zonal harmonics terms.
    /*! Can be set to 2, 4 or 6 (default). If "useGravity" is disabled, this setting
     * will have no effect.
     */
    int grav_jTermLevel;

    //! Variable to store the OPI Property "useAtmosphere".
    /*!
     * It states whether atmospherical perturbations should be taken into account
     * during propagation. Can be set to 0 (disabled) or 1 (enabled, default).
     */
    int opt_useAtmosphere;
```

```
//! Variable to store the OPI Property "useThirdBody".
/*!
 * It states whether third body perturbations should be taken into account during
 * propagation. Can be set to 0 (disabled) or 1 (enabled, default).
 */
int opt_useThirdBody;

//! Variable to store the OPI Property "useGravity".
/*!
 * It states whether gravitational perturbations should be taken into account
 * during propagation. Can be set to 0 (disabled) or 1 (enabled, default).
 */
int opt_useGravity;

//! Variable to store the OPI Property "useSolarRadiation".
/*!
 * It states whether solar radiation pressure perturbations should be taken into
 * account during propagation. Can be set to 0 (disabled) or 1 (enabled, default).
 */
int opt_useSolarRadiation;

//! Variable to store the OPI Property "cartesianPosition".
/*!
 * It states whether Ikebana should calculate the objects' Cartesian positions and
 * make them available through OPI. Can be set to 0 (disabled, default) or 1
 * (enabled).
 */
// Not to be confused with cardassianPosition which is somewhere near the Bajoran
// sector.
int opt_cartesianPosition;

//! Variable to store the OPI Property "verboseLevel".
/*!
 * It states how much output should be generated during program execution. Valid
 * settings range from 0 (default output) to 5 (lots of output); a negative value
 * will suppress all messages.
 */
int opt_verboseLevel;

//! An instance of the OPI::PerturbationModule handling unperturbed motion.
PMMeanMotion pMotion;
//! An instance of the OPI::PerturbationModule handling gravitational perturbations.
PMZonalHarmonics pGravity;
//! An instance of the OPI::PerturbationModule handling the third-body perturbations.
PMLuniSolar pThird;
//! An instance of the OPI::PerturbationModule handling atmospherical perturbations.
PMAtmosphere pAtmo;
//! An instance of the OPI::PerturbationModule handling solar radiation perturbations.
PMSolarRadiation pSolar;

//! An array of orbits storing the accumulated perturbations from each time step.
OPI::Orbit* deltaOrbit;
//! An array of orbits storing the orbital data given to Ikebana upon initialization.
OPI::Orbit* originalOrbit;
//! States the number of objects in the current Population.
int objectCount;

//! Checks the output of CUDA for errors.
/*!
 * In case of an error, program execution is terminated.
 */
void checkCUDAError();
//! Translates the given CUDA error code into a human-readable debug message.
/*!
 * In case of an error, program execution is terminated.
 */
```

```
    void checkCUDAError(cudaError_t err);
};

// CUDA functions currently have to be declared outside of classes.

// Calculates Cartesian position and velocity for a given orbit.
__device__ void calculateCartesianPosition(
  OPI::Orbit& orbit,
  OPI::Vector3& position,
  OPI::Vector3& velocity
);

// Resets the delta orbit and checks input values.
__device__ void preprocess(
  OPI::Orbit& orbit,
  OPI::Orbit& deltaOrbit,
  OPI::Orbit& originalOrbit,
  OPI::ObjectProperties& props
);

// Adds delta and original orbit to form new resulting orbit.
__device__ void postprocess(
  OPI::Orbit& orbit,
  OPI::Orbit& deltaOrbit,
  OPI::Orbit& originalOrbit,
  double julian_day
);


// Wrapper functions for the above kernels that set grid and block sizes.
__global__ void cartesianGPU(
  OPI::Orbit* orbit,
  OPI::Vector3* position,
  OPI::Vector3* velocity,
  int size
);

__global__ void preprocessGPU(
  OPI::Orbit* orbit,
  OPI::Orbit* deltaOrbit,
  OPI::Orbit* originalOrbit,
  OPI::ObjectProperties* props,
  int size
);

__global__ void postprocessGPU(
  OPI::Orbit* orbit,
  OPI::Orbit* deltaOrbit,
  OPI::Orbit* originalOrbit,
  double julian_day,
  int size
);

#define OPI_IMPLEMENT_CPP_PROPAGATOR Ikebana

#include "OPI/opi_implement_plugin.h"

#endif
```

C.2. Ikebana::PMMeanMotion

```
#ifndef __PERTURBATION_MEAN_MOTION_H__
#define __PERTURBATION_MEAN_MOTION_H__

#include "OPI/opi_cpp.h"

//! OPI::PerturbationModule that calculates the unperturbed motion of a satellite.
/*! This is the simplest perturbation module possible and can be used as a reference
 * to implement your own. Look into other modules for advanced stuff like configuration
 * via properties and multiple CUDA kernels.
 */
class PMMeanMotion: public OPI::PerturbationModule
{
  public:
    OPI::ErrorCode setTimeStep(double julian_date);

  protected:
    //! OPI interface function that calculates the unperturbed motion.
    /*! It simply calls the respective CUDA kernel that contains the
     *  actual equation. The resulting mean anomaly is stored in the
     *  delta orbit.
     */
    OPI::ErrorCode runCalculation(OPI::Population& data, OPI::Orbit* delta, float dt);

};

// CUDA functions currently have to be declared outside of classes.

// Kernel function that performs the actual calculation.
__global__ void calculateMotion(
  OPI::Orbit* orbit,
  OPI::Orbit* deltaOrbit,
  int size,
  float dt
);


#endif
```

C.3. Ikebana::PMZonalHarmonics

```
#ifndef __PERTURBATION_GRAVITY_H__
#define __PERTURBATION_GRAVITY_H__

#define OPI_CUDA_PREFIX __host__ __device__

#include "OPI/opi_cpp.h"
#include "OPI/opi_perturbation_module.h"

//! OPI::PerturbationModule containing zonal harmonics perturbations.
/*! These are analytical calculations based on Vallado (2007),
 * equations 9-38, 9-40 and 9-42.
 */
class PMZonalHarmonics: public OPI::PerturbationModule
{
  public:
    //! Class constructor.
    /*! Currently does nothing except setting the default jTerm level to six.
     */
    PMZonalHarmonics();

    //! Class destructor. Does nothing.
    ~PMZonalHarmonics();

    //! OPI interface function to set the current propagation time step.
    /*! Since zonal harmonics are independent of time, this function serves no
     * purpose here.
     */
    OPI::ErrorCode setTimeStep(double julian_date);

  protected:
    //! OPI interface function that calculates zonal harmonics perturbations.
    /*! In Ikebana, it calls the respective CUDA kernel which performs the actual
     * calculations.
     * @param data The OPI::Population containing the objects.
     * @param delta An array of OPI::Orbits to which the changes caused by this
     * perturbation module are added.
     * @param dt The propagation step size.
     * @return OPI::SUCCESS or an error message.
     */
    OPI::ErrorCode runCalculation(OPI::Population& data, OPI::Orbit* delta, float dt);

  private:
    //! Local variable for jTerm property. It can be set via the setProperty function.
    int jTermAccuracy;
};

// CUDA functions currently have to be declared outside of classes.

// Kernel that calculate zonal harmonics perturbations.
__device__ void gravity(
  OPI::Orbit& orbit,
  OPI::Orbit& deltaOrbit,
  float dt, int jTermAccuracy
);

// Wrapper function for the above kernel that sets block and grid size.
__global__ void gravityGPU(
  OPI::Orbit* orbit,
  OPI::Orbit* deltaOrbit,
  float dt, int jTermAccuracy, int size
);

#endif
```

C.4. Ikebana::PMLuniSolar

```
#ifndef __PERTURBATION_LUNI_SOLAR_H__
#define __PERTURBATION_LUNI_SOLAR_H__

#define OPI_CUDA_PREFIX __host__ __device__

#include "OPI/opi_cpp.h"

//! Struct to store relevant parameters of perturbing third bodies relative to the Earth.
/*! These are equatorial inclination, right ascension of ascending node, argument
 *  of mean longitude (called "u_3" in Vallado), gravitational constant, and distance
 *  to Earth in km.
 */
struct tThirdBody {
  float inclination;          //i_3, required to calculate direction
  float raan;             //Omega_3, required to calculate direction
  float arg_of_mean_longitude;  //u_3, required to calculate direction
  float gravitational_parameter;  //mue_3, required to calculate deltaOrbit
  float radius_to_earth;       //r_3, required to calculate deltaOrbit
};

//! OPI::PerturbationModule responsible for the calculation of Sun and Moon perturbations.
/*! Based on Cook's model and implemented as described in Vallado, chapter 9.6.3.
 */
class PMLuniSolar: public OPI::PerturbationModule
{
  public:
    //! Class constructor. Does nothing.
    PMLuniSolar();
    //! Class destructor. Does nothing.
    ~PMLuniSolar();

    //! OPI interface function to set the current Julian date prior to propagation.
    /*! In this module, this function is also used to calculate the parameters of
     * Sun and Moon based on the given date. Since these parameters are depent only
     * on the date and not on the properties of individual objects, they are
     * calculated once per time step on the CPU and then passed on to the GPU via the
     * kernel call.
     * param julian_date The date of the current time step.
     */
    OPI::ErrorCode setTimeStep(double julian_date);

  protected:
    //! OPI interface function to start calculating the perturbation.
    OPI::ErrorCode runCalculation(OPI::Population& data, OPI::Orbit* delta, float dt);

  private:
    //! Calculates relevant sun parameters based on the current date.
    tThirdBody getSunParameters(float julianDate);
    //! Calculates relevant moon parameters based on the current date.
    tThirdBody getMoonParameters(float julianDate);
    //! Stores results from getSunParameters
    tThirdBody sun;
    //! Stores results from getMoonParameters
    tThirdBody moon;
};


// CUDA functions currently have to be declared outside of classes.

// Calls calculateThirdBody for Sun and Moon and adds the results to the delta orbit.
__device__ void thirdbody(
  OPI::Orbit& orbit,
  OPI::Orbit& deltaOrbit,
```

```
  float dt,
  tThirdBody sun,
  tThirdBody moon
);

// Calculates the perturbations for the given third body.
__device__ OPI::Orbit calculateThirdBody(tThirdBody third, OPI::Orbit orbit);

// Wrapper function for the above kernel that sets block and grid size.
__global__ void thirdbodyGPU(
  OPI::Orbit* orbit,
  OPI::Orbit* deltaOrbit,
  float dt,
  int size,
  tThirdBody sun,
  tThirdBody moon
);

#endif
```

C.5. Ikebana::PMSolarRadiation

```
#ifndef __PERTURBATION_SOLAR_RADIATION_H__
#define __PERTURBATION_SOLAR_RADIATION_H__

#define OPI_CUDA_PREFIX __host__ __device__

#include "OPI/opi_cpp.h"

//! Struct that stores two angles for Escobal's shadow function.
struct tPerifocal {
  float beta;
  float xi;
};

//! Struct that defines the necessary properties of the Earth's shadow.
struct tShadow {
  float trueAnomalyEntry;       // true anomaly of shadow entry point
  float trueAnomalyExit;        // true anomaly of shadow exit point
  float eccentricAnomalyEntry;  // ecc. anomaly of shadow entry point
  float eccentricAnomalyExit;   // ecc. anomaly of shadow exit point
  float radiusEntry;            // radius vector to entry point
  float radiusExit;             // radius vector to exit point
  bool crossed;                 // true if the shadow was crossed
};

//! Struct that defines the acceleration vector of the solar radiation pressure.
struct tAcceleration {
  float r;
  float s;
  float wsin;
  float wcos;
};


//! OPI::PerturbationModule responsible for the calculation of SRP perturbations.
/*! This module calculates solar radiation pressure perturbations based on an
 * analytical model by Escobal (1965).
 */
class PMSolarRadiation: public OPI::PerturbationModule
{
  public:
    //! OPI interface function to set the current Julian date prior to propagation.
    /*! Calculates the Sun's position at the given time. This vector, as well as
     * two intermediate results, obliquity and ecliptic longitude, are stored
     * as class attributes for later upload to the GPU.
     * param julian_date The date of the current time step.
     */
    OPI::ErrorCode setTimeStep(double julian_date);

  protected:
    //! OPI interface function to start calculating the perturbation.
    OPI::ErrorCode runCalculation(OPI::Population& data, OPI::Orbit* delta, float dt);

  private:
    //! Calculates the ecliptic longitude of the Sun at the given date.
    float getSunEclipticLongitude(double julianDate);

    //! Axial tilt of the Earth at the current time
    float obliquity;
    //! Ecliptic longitude of the Sun at the current time
    float sunEclipticLongitude;
    //! Position of the Sun at the current time
    OPI::Vector3 sunPosition;
};
```

```
// CUDA functions currently have to be declared outside of classes.


// Calculates the shadow properties for the given orbit and Sun position.
__device__ tShadow calculateShadow(OPI::Orbit orbit, OPI::Vector3 sunPosition);

// Calculates the angles beta and xi required for the shadow function.
__device__ tPerifocal calculatePerifocal(OPI::Orbit orbit, OPI::Vector3 sunPosition);

// Calculates Escobal's shadow function.
__device__ float shadowFunction(OPI::Orbit orbit, tPerifocal p, float trueAnomaly);

// Calculates shadow entry and exit points with the help of a quartic equation.
__device__ tShadow shadowEntryExit(
  double roots[4],
  int nroots,
  tPerifocal p,
  OPI::Orbit orbit
);

// Calculates the perturbance of the given shadow on the given orbit.
__device__ void calculateShadowInfluence
(
  OPI::Orbit orbit,
  tShadow shadow,
  tAcceleration SRPAcceleration,
  OPI::Orbit& deltaOrbit,
  float dt
);

// Calculates the acceleration vector of the solar radiation pressure.
__device__ tAcceleration getSRPAcceleration(
  OPI::Orbit orbit,
  OPI::ObjectProperties props,
  float obliquity,
  float sunEclipticLongitude
);

// Main kernel function that calculates the SRP perturbance.
__device__ void solar(
  OPI::Orbit& orbit,
  OPI::ObjectProperties& props,
  OPI::Orbit& deltaOrbit,
  float obliquity,
  OPI::Vector3 sunPosition,
  float sunEclipticLongitude,
  float dt
);

// Wrapper function for the above kernel that sets block and grid size.
__global__ void solarGPU(
  OPI::Orbit* orbit,
  OPI::ObjectProperties* props,
  OPI::Orbit* deltaOrbit,
  int size,
  float obliquity,
  OPI::Vector3 sunPosition,
  float sunEclipticLongitude,
  float dt
);

#endif
```

C.6. Ikebana::AtmosphericData

```
#ifndef __PERTURBATION_ATMOSPHERE_DATA_H__
#define __PERTURBATION_ATMOSPHERE_DATA_H__

#include "fileManager.h"
#include "AtmosphericDataTypes.h"

#include <map>
#include <iostream>
#include <string>

using namespace std;

//! Data provider for Ikebana's atmospheric model
class AtmosphericData
{
  public:
    //! Class constructor. Does nothing.
    AtmosphericData();

    //! Class destructor. Does nothing.
    ~AtmosphericData();

    //! Initialization function that is called by PMAtmosphere.
    /*! This function loads the files atmosphere.tab, solaractivity.txt and
     *  bessel_interpolated.dat which usually reside inside the archive ikebana.dat.
     *  Actual file loading is handled by the FileManager.
     *  Content is stored in RAM (solar activity) or CUDA unified memory
     *  (atmosphere, bessel).
     *  return true if initialization was completed without errors, false otherwise.
     */
    bool init();

    //! Called upon disabling the atmospheric module. Cleans up allocated memory.
    /*! return true if cleanup was completed successfully, false otherwise.
     */
    bool cleanup();

    //! Calculates the indices for the given day's Ap, mean and daily F10.7.
    /*! Called by the setTimeStep function of PMAtmosphere.
     *  The results are stored as unified memory pointers so the CUDA threads can
     *  look them up. These pointers are returned by the getOffsetPointer function.
     *  param julianDay The current date
     */
    void calculateOffsets(float julianDay);

    //! Returns a pointer to the atmospheric data table in device memory.
    tAtmoData* getDataPointer();

    //! Returns a pointer to an int[3] containing ap, mean and daily F10.7 indices.
    int* getOffsetPointer();

    //! Returns a pointer to the Bessel function lookup table in unified memory.
    double* getBesselPointer();

  private:
    //! Instance of FileManager used to load data.
    IkebanaFileManager fm;

    //! Map to store data solar activity data.
    /*! The index is the Julian Date, rounded to days.
     */
    map<int, tSolarActivity> solar;
```

```
    //! Atmospherical data stored in device memory.
    /*! getDataPointer() returns a pointer to it.
     */
    tAtmoData* atmo;

    //! Indices for ap and f10.7 values into atmosphere.tab.
    /*! getOffsetPointer() returns a pointer to it.
     */
    int* currentOffsets;

    //! Array to store the lookup table for the Bessel function.
    /*! getBesselPointer() returns a pointer to it.
     */
    double* besselLookup;

    //! Loads atmosphere.tab and stores it in device memory.
    void loadAtmosphere();

    //! Loads solaractivity.txt and stores it in unified memory.
    void loadSolarActivity();

    //! Loads Bessel function lookup table and stores it in unified memory.
    void loadBesselLookup();

    //! Finds the next index for a value within a given range and step size.
    /*! This function is used to find the correct indices for the atmospheric table.
     *  For example, (14, 0, 60, 15) would return 15 because it's the next value in
     *  the sequence "0 15 30 45 60".
     *  param value The original value for which a match is sought. If this value is
     *     outside the range, min or max are returned depending on which is closest.
     *  param min The minimum value of the sequence.
     *  param max The maximum value of the sequence.
     *  param stepSize The step size of the sequence.
     *  return Next higher match.
     */
    int findNext(int value, int min, int max, int stepSize);

#if __cplusplus <= 201103L
    // replacements for string conversion functions that are only available in C++11
    inline double stod(string s) { return strtod(s.c_str(),NULL); };
    inline float stof(string s) { return atof(s.c_str()); };
    inline int stoi(string s) { return atoi(s.c_str()); };
#endif


};


// CUDA functions currently have to be declared outside of classes.

// Called by PMAtmosphere to get interpolated lookup table data at the given location
// with the given input values.
__device__ tAtmoData getDataAt(int km, tAtmoData* atmoPointer, int* offsetPointer);

// generate atmospheric data by interpolating the values around the given indices
__device__ tAtmoData trilinearInterpolation(
  int dailyf107,
  int ap,
  int meanf107,
  int km,
  tAtmoData* atmoPointer
);

// CUDA equivalent of the above findNext() function.
__device__ int findNextGPU(int value, int minValue, int maxValue, int stepSize);
```

```
// Fetches a point from the atmospheric data table. Called during interpolation.
__device__ tAtmoData getPointAt(
  int dailyf107,
  int ap,
  int meanf107,
  int km,
  tAtmoData* atmoPointer
);

// Fetches an interpolated value from the Bessel function lookup table.
__device__ double getBesselValue(int order, int value, double* besselPointer);


#endif
```

C.7. Ikebana::PMAtmosphere

```
#ifndef _PERTURBATION_ATMOSPHERE_H_
#define _PERTURBATION_ATMOSPHERE_H_

#define OPI_CUDA_PREFIX __host__ __device__

#include "OPI/opi_cpp.h"
#include "AtmosphericData.h"
#include "AstroMathCUDA.h"

//! OPI::PerturbationModule responsible for the calculation of atmospheric drag.
/*! Receives NRLMSISE-00 data from AtmosphericData and uses analytical models
 * published by King-Hele (1987).
 */
class PMAtmosphere: public OPI::PerturbationModule
{
  public:
    //! Class constructor. Used to initialize some variables.
    PMAtmosphere();
    //! Class destructor. Does nothing.
    ~PMAtmosphere();
    //! Calls the init function of AtmosphericData which loads the lookup tables.
    void init();
    //! Calls the cleanup function of AtmosphericData which frees memory.
    void cleanup();
    //! Returns true if the module has been initialized via the init function.
    bool isInitialized();

    //! OPI interface function to set the current Julian date prior to propagation.
    /*! In this module, the function calls the calculateOffsets function of
     * AtmosphericData to determine the time-dependent offsets into the atmospheric
     * data table and upload them to the GPU.
     * param julian_date The date of the current time step.
     */
    OPI::ErrorCode setTimeStep(double julian_day);

  protected:
    //! OPI interface function to start calculating the perturbation.
    OPI::ErrorCode runCalculation(OPI::Population& data, OPI::Orbit* delta, float dt);

  private:
    //! Attribute to store the initialization status. Set to false in the constructor.
    bool initialized;
    //! Instance of AtmosphericData that provides access to the lookup tables.
    AtmosphericData dataProvider;
    //! The current Julian date is stored as an attribute for later GPU upload.
    float julianDay;
};


// CUDA functions currently have to be declared outside of classes.

// Auxiliary function for King-Hele's equations in high-eccentricity cases.
__device__ double hele(double eccentricity);

// Auxiliary function for King-Hele's equations in high-eccentricity cases.
__device__ double findroot(double c, float eccentricity);

// Correction function to compensate the simplifications of the density lookup table.
__device__ tAtmoData getDensityFactorsAt(float altitude, float julianDay, float raop);

// Solves the Bessel function based on a lookup table in AtmosphericData.
__device__ double bessel(int order, float coeff, double* besselPointer);
```

```
// Calculates the atmospheric perturbation.
__device__ void atmosphere(
  OPI::Orbit& orbit,
  OPI::ObjectProperties& props,
  OPI::Orbit& deltaOrbit,
  tAtmoData* dataPointer,
  int* offsetPointer,
  double* besselPointer,
  float julianDay,
  float dt
);

// Wrapper function for the above kernel that sets block and grid size.
__global__ void atmosphereGPU(
  OPI::Orbit* orbit,
  OPI::ObjectProperties* props,
  OPI::Orbit* deltaOrbit,
  int size,
  tAtmoData* dataPointer,
  int* offsetPointer,
  double* besselPointer,
  float julianDay,
  float dt
);

#endif
```

Bibliography

[Ahn, 2012] Ahn, M. (August 2012). GPU Accelerated Satellite Orbit Propagation. `https://github.com/ahnm/cis565-project`. Accessed 2015-04-10.

[Akenine-Möller and Johnsson, 2012] Akenine-Möller, T. and Johnsson, B. (2012). Performance per what? *Journal of Computer Graphics Techniques (JCGT)*, 1(1):37–41.

[Amdahl, 1967] Amdahl, G. M. (1967). Validity of the single processor approach to achieving large scale computing capabilities. In *Proceedings of the April 18-20, 1967, Spring Joint Computer Conference*, AFIPS '67 (Spring), pages 483–485. ACM.

[Blachford, 2006] Blachford, N. (2006). Lets Get Physical: Inside The PhysX Physics Processor. `http://www.blachford.info/computer/articles/PhysX1.html`. Accessed 2015-07-14.

[Bowman, 2002] Bowman, B. R. (2002). True Satellite Ballistic Coefficient Determination for HASDM. *AIAA/AAS Astrodynamics Specialist Conference and Exhibit.*

[Bruinsma et al., 2012] Bruinsma, S. L., Sanchez-Ortiz, N., Olmedo, E., and Guijarro, N. (2012). Evaluation of the DTM-2009 thermosphere model for benchmarking purposes. *Journal of Space Weather and Space Climate.*

[Bundeskunsthalle, 2014] Bundeskunsthalle (2014). *OUTER SPACE. Faszination Weltraum.* nicolai Verlag.

[Buschmann et al., 1996] Buschmann, F., Meunier, R., Rohnert, H., Sommerlad, P., and Stal, M. (1996). *Pattern-Oriented Software Architecture. Volume 1.* Wiley.

[CCSDS, 2009] CCSDS (2009). Consultive Committee for Space Data System - Orbit Data Messages - Recommended Standard.

[CCSDS, 2010] CCSDS (2010). Consultive Committee for Space Data System - Navigation Data - Definitions and Conventions.

[Cook, 1962] Cook, G. E. (1962). Luni-Solar Perturbations of the Orbit of an Earth Satellite. *Geophysical Journal of the Royal Astronomical Society.*

[d'Eon et al., 2007] d'Eon, E., Luebke, D., and Enderton, E. (2007). Efficient Rendering of Human Skin. *Eurographics Symposium on Rendering.*

[Escobal, 1965] Escobal, P. (1965). *Methods of Orbit Determination.* Krieger Publishing, Inc.

[Flegel, 2007] Flegel, S. (2007). Simulation der zukünftigen Population von Raumfahrtrückständen unter Berücksichtigung von Vermeidungsmaßnahmen. Master's thesis, Technische Universität Carolo-Wilhelmina zu Braunschweig.

[Flegel, 2013] Flegel, S. (2013). *Multi-Layer Insulation as Contribution to Orbital Debris.* PhD thesis, Technische Universität Carolo-Wilhelmina zu Braunschweig.

[Flegel et al., 2011] Flegel, S., Gelhaus, J., Möckel, M., Wiedemann, C., Kempf, D., and Krag, H. (2011). Maintenance of the ESA MASTER Model - Final Report, Revision 1.1. Technical report, ESA/ESOC.

[Flegel et al., 2010] Flegel, S., Möckel, M., Gelhaus, J., Wiedemann, C., and Vörsmann, P. (2010). *Analyse zur Deutschen Position zur Wirtschaftlichkeit von Space Debris-Vermeidungsmaßnahmen.*

[Fraire et al., 2013] Fraire, J., Ferreyra, P., and Marques, C. (2013). OpenCL-Accelerated Simplified General Perturbations 4 Algorithm. *Proceedings of the 14th Argentine Symposium of Technology.*

[Gessler et al., 2014] Gessler, A., Schulze, T., Kulling, K., and Nadlinger, D. (2014). Open Asset Import Library. `http://assimp.sourceforge.net`. Accessed 2015-08-09.

[Gordon, 2010] Gordon, R. C. (2010). PhysicsFS API Documentation. `https://icculus.org/physfs/docs/html`. Accessed 2014-05-08.

[Harrison and Waldron, 2007] Harrison, O. and Waldron, J. (2007). AES Encryption Implementation and Analysis on Commodity Graphics Processing Units. *Cryptographic Hardware and Embedded Systems - CHES 2007.*

[Hobson and Clarkson, 2012] Hobson, T. A. and Clarkson, V. L. (2012). GPU-based Space Situational Awareness Simulation utilising parallelism for enhanced multi-sensor management.

[Ierusalimschy et al., 2006] Ierusalimschy, R., de Figueiredo, L. H., and Celes, W. (2006). *Lua 5.1 Reference Manual.* Lua.org.

[Kebschull, 2011] Kebschull, C. (2011). Parallelisierung eines numerischen Propagators mit OpenCL. Master's thesis, Technische Universität Carolo-Wilhelmina zu Braunschweig.

[Kelso, 2000] Kelso, T. S. (2000). Space Weather Data. `http://celestrak.com/SpaceData/SpaceWx-format.asp`. Accessed 2015-07-14.

[Kent, 2004] Kent, S. L. (2004). *The Making of Doom III.* McGraw-Hill/Osborne.

[Kessler and Cour-Palais, 1978] Kessler, D. J. and Cour-Palais, B. G. (1978). Collision frequency of artificial satellites: The creation of a debris belt. *Journal of Geophysical Research, Volume 83, Issue A6, p. 2637-2646*, pages 2637–2646.

[Khronos OpenCL Working Group, 2015] Khronos OpenCL Working Group (2015). *The OpenCL Specification Version 2.1, Revision 8.*

[King-Hele, 1987] King-Hele, D. (1987). *Satellite Orbits in an Atmosphere.* Blackie and Son Ltd.

[Kirk and Hwu, 2010] Kirk, D. B. and Hwu, W. W. (2010). *Programming Massively Parallel Processors - A Hands-on Approach.* Elsevier.

[Klinkrad, 2006] Klinkrad, H. (2006). *Space Debris - Models and Risk Analysis*. Springer.

[Köhncke, 2014] Köhncke, M. (2014). Automatisierter Vergleich von Bahnpropagatoren. Master's thesis, Technische Universität Carolo-Wilhelmina zu Braunschweig.

[Latinga, 2014] Latinga, S. (2014). Simple DirectMedia Layer. `https://www.libsdl.org`. Accessed 2015-08-02.

[Liou, 2006] Liou, J.-C. (2006). Collision activities in the future orbital debris environment. *Advances in Space Research, Vol. 38*.

[Lorefice, 2010] Lorefice, D. (2010). Scientific Parallel Animation and Computing Environment - Simulation. Master's thesis, Technische Universität Carolo-Wilhelmina zu Braunschweig.

[Margolin, 2001] Margolin, J. (2001). The Secret Life of Vector Generators. `http://www.jmargolin.com/vgens/vgens.htm`. Accessed 2015-04-15.

[Marquardt, 1999] Marquardt, K. (1999). Patterns for Plug-Ins. *Proceedings of EuroPLoP 1999*.

[MIT Technology Review, 2002] MIT Technology Review (2002). Innovators under 35. `http://www2.technologyreview.com/tr35/profile.aspx?trid=243`. Accessed 2015-04-10.

[Möckel et al., 2013] Möckel, M., Flegel, S., Kebschull, C., Braun, V., Miller, A., Gelhaus, J., Wiedemann, C., Vörsmann, P., and Kreisel, J. (2013). *SD-LEO - Wirtschaftlichkeit der Stabilisierung der Space Debris-Population auf niedrigen Erdumlaufbahnen*.

[Möckel et al., 2012] Möckel, M., Kebschull, C., Flegel, S., Gelhaus, J., Braun, V., Wiedemann, C., and Vörsmann, P. (2012). Flexible Implementation of Orbital Propagators in Heterogenous Computing Environments. *DGLR-Jahrestagung 2012, Berlin*.

[Möckel et al., 2015] Möckel, M., Radtke, J., Wiedemann, C., and Stoll, E. (2015). *ELA - Erweiterte Langzeitanalyse der zukünftigen Weltraummüllpopulation unter Berücksichtigung aktiver Entfernungsmaßnahmen*.

[Möckel et al., 2011] Möckel, M., Wiedemann, C., Flegel, S., Gelhaus, J., Klinkrad, H., Krag, H., and Vörsmann, P. (2011). Using Parallel Computing for the Display and Simulation of the Space Debris Environment. *Advances in Space Research 48, pp. 173-183*.

[Najjar, 2014] Najjar, M. (2014). *outer space*. DISTANZ Verlag.

[NVIDIA Corporation, 2009] NVIDIA Corporation (2009). *NVIDIA's Next Generation CUDA Compute Architecture*.

[NVIDIA Corporation, 2013] NVIDIA Corporation (2013). *NVIDIA Tesla - Kepler Family Datasheet*.

[NVIDIA Corporation, 2015] NVIDIA Corporation (2015). *CUDA C Programming Guide, Version 7.0*.

[OpenACC Group, 2013] OpenACC Group (2013). *The OpenACC Application Programming Interface, Version 2.0*.

[OpenMP Architecture Review Board, 2013] OpenMP Architecture Review Board (2013). *OpenMP Application Program Interface, Version 4.0.*

[Owens et al., 2008] Owens, J. D., Houston, M., Luebke, D., Green, S., Stone, J. E., and Phillips, J. C. (2008). GPU Computing. *Proceedings of the IEEE, Vol. 96, No. 5.*

[Peus, 2013] Peus, C. (May 2013). Weltansichten / World Views. *art.*

[Picone et al., 2002] Picone, J., Hedin, A., Drob, D., and Aikin, A. (2002). NRL-MSISE-00 Empirical Model of the Atmosphere: Statistical Comparisons and Scientific Issues. *J. Geophys. Res., doi:10.1029/2002JA009430.*

[Radtke, 2011] Radtke, J. (2011). Modellieren der Erdatmosphäre zur Bahnlebensdauerberechnung erdgebundener Objekte. Master's thesis, Technische Universität Carolo-Wilhelmina zu Braunschweig.

[Reglitz, 2012] Reglitz, S. (2012). Architektur eines Tools zur Vorhersage von Objektbahnen. Master's thesis, Technische Universität Carolo-Wilhelmina zu Braunschweig.

[Rodermund, 2010] Rodermund, T. (2010). Scientific Parallel Animation and Computing Environment - Visualization. Master's thesis, Technische Universität Carolo-Wilhelmina zu Braunschweig.

[Rohrbeck, 2011] Rohrbeck, M. (2011). Validierung der West Ford Needles - Simulation. Master's thesis, Technische Universität Carolo-Wilhelmina zu Braunschweig.

[Space-Track.org, 2015] Space-Track.org (2015). TLE Full Catalog. `https://www.space-track.org/basicspacedata/query/class/tle_latest/ORDINAL/1/EPOCH/>now-30/orderby/NORAD_CAT_ID/format/tle`. Login required. Accessed 2015-05-11, last NORAD ID: 40640.

[Sutter and Larus, 2005] Sutter, H. and Larus, J. (2005). Software and the Concurrency Revolution. *ACM Queue.*

[Szirmay-Kalos et al., 2010] Szirmay-Kalos, L., Umenhoffer, T., Tóth, B., Szécsi, L., , and Sbert, M. (2010). Volumetric Ambient Occlusion. Technical report, Technical University of Budapest.

[Tapping and Charrois, 1993] Tapping, K. and Charrois, D. (1993). Limits to the Accuracy of the 10.7cm Flux. *Solar Physics.*

[Thomsen, 2013] Thomsen, P. (2013). GPU-Basierte Analyse von Kollisionen im Weltraum. Master's thesis, Technische Universität Carolo-Wilhelmina zu Braunschweig.

[Thomsen and Möckel, 2013] Thomsen, P. and Möckel, M. (2013). OPI Source Code on GitHub. `https://github.com/ILR/OPI`. Accessed 2015-08-19.

[Truelsen, 2007] Truelsen, R. (2007). Real-time Shallow Water Simulation and Environment Mapping and Clouds. Technical report, University of Copenhagen.

[Turner, 2014] Turner, P. D. (2014). CEGUI Developer Documentation. `http://static.cegui.org.uk/docs/current`. Accessed 2015-08-02.

[Vallado et al., 2006] Vallado, D., Crawford, P., Hujsak, R., and Kelso, T. (2006). Revisiting Spacetrack Report #3. *AIAA 2006-6753*.

[Vallado, 2007] Vallado, D. A. (2007). *Fundamentals of Astrodynamics and Applications*. Microcosm Press, 3rd edition.

[van Waveren, 2013] van Waveren, J. P. (2013). DOOM 3 BFG Technical Note. Technical report, id Software.

[Wiedemann, 2014] Wiedemann, C. (2014). Raumfahrttechnische Grundlagen.

[Zuschlag, 1985] Zuschlag, J. (1985). Programmierung der direkten numerischen Integration der gestörten Bewegungsgleichung eines Satelliten. Master's thesis, Technische Universität Carolo-Wilhelmina zu Braunschweig.